MW00478882

A Catholic Scientist Proves God Exists

Also by Gerard M. Verschuuren
from Sophia Institute Press:

Forty Anti-Catholic Lies
A Mythbusting Apologist
Sets the Record Straight

In the Beginning
A Catholic Scientist Explains
How God Made Earth Our Home

Dr. Gerard M. Verschuuren

A Catholic Scientist
Proves God Exists

SOPHIA INSTITUTE PRESS
Manchester, New Hampshire

Copyright © 2019 by Gerard M. Verschuuren

Printed in the United States of America. All rights reserved.

Cover by David Ferris Design.

Cover image: vector design (275517830) © Marina Sun / Shutterstock.

Unless otherwise noted, Scripture quotations in this work are taken from the *New American Bible, revised edition* © 2010, 1991, 1986, 1970 Confraternity of Christian Doctrine, Washington, D.C., and are used by permission of the copyright owner. All Rights Reserved. No part of the *New American Bible* may be reproduced in any form without permission in writing from the copyright owner. Quotations marked "RSVCE" are taken from the *Revised Standard Version of the Bible: Catholic Edition*, copyright © 1965, 1966 the Division of Christian Education of the National Council of the Churches of Christ in the United States of America. Used by permission. All rights reserved.

Excerpts from the English translation of the *Catechism of the Catholic Church*, second edition, © 1994, 1997, 2000 by Libreria Editrice Vaticana–United States Conference of Catholic Bishops, Washington, D.C. All rights reserved.

No part of this book may be reproduced, stored in a retrieval system, or transmitted in any form, or by any means, electronic, mechanical, photocopying, or otherwise, without the prior written permission of the publisher, except by a reviewer, who may quote brief passages in a review.

Sophia Institute Press
Box 5284, Manchester, NH 03108
1-800-888-9344

www.SophiaInstitute.com

Sophia Institute Press® is a registered trademark of Sophia Institute.

Paperback ISBN 978-1-64413-104-6
eBook ISBN 978-1-64413-105-3
Library of Congress Control Number:2019957500

First printing

To the late Fr. James V. Schall, S.J.,
who guided many students at Georgetown University
on the way to God through philosophy

Contents

Introduction

Does God exist? That's the million-dollar question haunting the minds of almost everyone these days, whether or not we believe that God exists. Of course, it's not so much the question that counts but the answer. Where can we find an answer that everyone would accept? Could it be in science?

We will discuss first whether science can give us a reliable answer to the question of God's existence. If science can't give us that answer, as we will argue, then we must look somewhere else — preferably in a place that everyone has access to. Is there such a place? Yes, it's the domain of *reason*. Each one of us is a *reasonable*, or *rational*, being — traditionally called by Aristotle *animal rationale*. The so-called proofs of God's existence belong to the domain of reason and are therefore accessible to any reasonable, or rational, being.

We will carefully analyze these proofs of God's existence and correct the misconceptions many people have about them. We will then come to the conclusion that these proofs have more power than any scientific proof could ever bolster. So, does God exist? The resounding answer will be yes!

This book has a question-and-answer format. The questions come from someone — perhaps someone like you — who has critical questions about all that Christianity proclaims about God. The answers come from me, a scientist who is also a devoted Catholic. They are not exclusively my answers; they come from something much, much larger than me—the Catholic Church. Please don't assume ahead of time that these answers are suspicious, for the Catholic Church has always affirmed the role of reason in defending her faith. You don't have to be a Catholic to understand these answers, for they are accessible to anyone who believes in the power of reason. Just try it!

A Catholic Scientist Proves God Exists

1

Can Science Prove God's Existence?

Does God exist? Nowadays many people would say a reliable answer to this question can be found only through science. In fact, science has become the most respected source for providing us with reliable answers to any questions we have. True, science certainly has an impressive track record. If it has given us so many reliable answers to so many questions, why shouldn't we expect science to answer the question of God's existence?

The answer is not so obvious. Has science really given us answers to *all* our questions? Indeed, there is much that science can do, but that obscures the fact that there is also much that science cannot do. Most likely, we don't have to explain how much science can do for us — we see the results all around us, in astronomy, physics, chemistry, and genetics. But it probably requires more time to explain what science is *not* able to do for us.

To begin with, the very question as to whether science is capable of answering all our questions must be one of those questions that science is supposedly capable of answering. But there is no way science, on its own, can prove that it has an answer to all our questions. This would be possible only if we know all the possible questions people might have and all the answers that science has and will come up with. That's a task without end. One could agree with the late Nobel laureate and biologist Konrad Lorenz that a scientist "knows more and more about less and less and finally knows everything about nothing."[1] Science on its own is not able to prove that it has answers to anything we might ask.

Second, the idea that science is the only way of having all our questions answered is not scientific in itself. How could science ever prove, all by itself, that it is the only way of finding truth? There is no experiment that could do the trick. We cannot test this idea in the laboratory or with a full-roof experiment or a double-blind trial. The truth of the matter is that science, on its own, cannot answer questions that are beyond its scope and beyond the reach of its empirical and experimental techniques—unless we declare such questions invalid ahead of time. But that's a self-serving way out.

Third, the idea that any method as successful as the one science provides would disqualify any other methods is obviously an assumption, not a conclusion. A blood test, for instance, cannot be used to prove that a blood test is the best, let alone the only, method there is. Or take a metal detector: it's a perfect tool for detecting metals, but that does not make it a perfect tool for detecting anything else. There is no way metal detectors can tell us anything about the existence of stones or plants, let alone about nonmaterial entities such as God. An instrument can detect only what it is designed to detect. And the same is true of science. It cannot detect things for which it has no tools.

Fourth, saying that the territory of science covers all there is implies that there is nothing outside of that territory. Such a statement can be made only if one is able to step outside the domain of science to prove that there is nothing outside that domain. Obviously, this claim cannot be tested with tools and methods from

---

[1] Quoted in Larry Collins and Thomas Schneid, *Physical Hazards of the Workplace* (Boca Raton, FL: CRC Press, 2001), 107.

inside the domain of science. Simply put, science cannot pull itself up by its own bootstraps—any more than an electric generator is able to run on its own power.

Fifth, science has a rather narrow scope, restricted to what can be measured, dissected, counted, and quantified. But therein also lies its limitation. First, it limits itself exclusively to what can be measured, dissected, counted, and quantified, and then it declares anything else as nonexistent because it cannot be measured, dissected, counted, and quantified. That is a form of circular reasoning. Instead, we should tell ourselves differently: "Not everything that can be counted counts; not everything that counts can be counted."[2] The Austrian physicist and Nobel laureate Erwin Schrödinger said about science:

> It knows nothing of beautiful and ugly, good or bad, God and eternity. Science sometimes pretends to answer questions in these domains, but the answers are very often so silly that we are not inclined to take them seriously.[3]

Sixth, interestingly enough, the astonishing successes of science have not been gained by answering any kind of question, but precisely by refusing to do so. We shouldn't forget that science has purchased success at the cost of limiting its ambition. It has been doing so for centuries. However, anything we neglect we cannot therefore reject. Everything that cannot be dissected, measured, counted, or quantified is off-limits for science. That makes science

---

[2]   It is often said that this quote was written on a sign or a blackboard in Albert Einstein's office at Princeton University, but that is unsubstantiated. The quote is probably more recent and from American author and humorist William Bruce Cameron.

[3]   Erwin Schrödinger, *Nature and the Greeks* in *Nature and the Greeks and Science and Humanism* (Cambridge, MA: Cambridge University Press, 1964), 96.

"blind" for many aspects of life that are not material and therefore must be beyond the reach of science. As it has been famously said, "Gravitation is not responsible for people falling in love."[4]

Seventh, in spite of the fact that science is exclusively about material things, scientific research requires nonmaterial things, such as logic and mathematics. Logic and mathematics are not physical and therefore are not testable by the natural sciences—yet they cannot be ignored, let alone be denied, by science. In fact, science relies heavily on logic and mathematics to interpret the data that scientific observation and experimentation provide. Yet these nonmaterial things are real and indispensable, even though they are beyond scientific observation. Take, for instance, the mathematical concept of *pi* (π). It is not a material object like a melon; it is not even a property of material things, for there are no *pi*-sided melons. Instead, *pi* is a precise and definite concept with logical and mathematical relationships to other equally precise concepts, and therefore it can be helpful in science.

Eighth, ironically, the idea that science is capable of answering all our questions is itself one of those nonmaterial things that science must, by definition, reject. This idea cannot be dissected, counted, measured, or quantified—and therefore it is not a scientific notion and cannot be tested by science. Those who claim that there must be a scientific answer to all our questions cannot declare their claim as a scientific one. It's at best an unfounded, dubious assumption.

Ninth, the idea that science is capable of answering all our questions makes scientists easily forget that they went through

---

[4] This quote is usually attributed to Isaac Newton, but also to Albert Einstein, jotted in German in the margin of a letter to him. See Helen Dukas, ed., *Einstein: The Human Side* (Princeton: Princeton University Press, 1981), 56.

hyperspecialized training in a narrow field of science coupled with a lack of exposure to other disciplines and methods. If they claim expertise in everything else as well, they are like plumbers who also try to fix your electricity or like electricians who attempt to fix your plumbing. What gets lost in the process is the awareness that expertise in one field may not help much in handling problems in another field. The late American physicist John A. Wheeler, who coined the term *black hole*, put it this way, "We live on an island surrounded by a sea of ignorance. As our island of knowledge grows, so does the shore of our ignorance."[5] This means that those who think they know everything and have answers for everything must live on a really tiny island.

Tenth, and probably most damaging, is the fact that scientists must hold many *assumptions*, which they may not want to acknowledge or are simply not aware of. These assumptions come *before* science can get started. They must assume that there is an objective world external to their minds. They must assume that this world is governed by regularities captured in scientific laws. They must assume that their methodology can uncover and accurately describe these regularities. Since their scientific methodology presupposes these assumptions, it cannot attempt to justify them in a scientific way without arguing in a circle. To break out of this circle, one must take the position of an extrascientific vantage point. The very existence of that extrascientific vantage point, however, would falsify the claim that science is the *only* way of investigating objective reality.[6] Science can never explain what science itself must take for granted.

5   Quoted in Clifford A. Pickover, *Wonders of Numbers* (New York: Oxford University Press, 2000), 195.
6   Edward Feser makes this point convincingly in *The Five Proofs of God's Existence* (San Francisco: Ignatius Press, 2017), 274.

A Catholic Scientist Proves God Exists

After all we have said so far—and there is probably much more to say—we must come to the conclusion that science can tell us something about only *one* aspect of the world: its scientific, material aspect. But that doesn't preclude the existence of other aspects and views. That's why the late University of California at Berkeley philosopher of science Paul Feyerabend could say: "Science should be taught as one view among many and not as the one and only road to truth and reality."[7] Even the "positivistic" philosopher Gilbert Ryle expressed a similar view: "The nuclear physicist, the theologian, the historian, the lyric poet and the man in the street produce very different, yet compatible and even complementary pictures of one and the same 'world.'"[8] Science provides only one of these pictures.

It's true that science has a great track record—but so does Hollywood. Having a great track record does not make something worth *everything*. Nevertheless, it remains an ongoing temptation to make extravagant claims about science. Science and scientific knowledge have thus been endowed with the status of a new, comprehensive philosophy of life, coming close to a new religion. The "building" that houses science has become a "temple" of science, which is still manifest in the way the Massachusetts Institute of Technology (MIT) was built in the late 1800s—the dream temple of a new "religion."

What you said so far about the limitations of science may be true in general, but that does not mean the door is wide open now to prove the existence of God. Shouldn't our knowledge at least be grounded on sense experience? Otherwise it's no more

---

[7] Paul Feyerabend, *Against Method: Outline of an Anarchistic Theory of Knowledge* (New York: Verso Books, 1975), viii.

[8] Gilbert Ryle, *Dilemmas* (Cambridge: Cambridge University Press, 1960), 68–81.

possible to prove the existence of God than to prove the existence of elves and fairies.

You are right. We do have a problem here—at least at first sight. It's the problem of God eluding any scientific exploration. There are, however, several reasons why God is not and cannot be known through sense experience and why we cannot test for God in the laboratory. Let's discuss these reasons briefly.

The first reason is that science limits itself to material entities and phenomena—things that can be measured, counted, quantified, and dissected. But because God is not a material entity, as we will explain later in the book, the technique of dissecting, counting, measuring, and quantifying does not and cannot work when it comes to God. By limiting itself this way, science can no longer answer questions about God. But that doesn't mean science has proven that God does *not* exist. You cannot merely reject what you neglect.

The second reason is that we should be careful to demand indisputable evidence for God's existence by testing for God in the laboratory. Catholic philosophers—such as Peter Kreeft of Boston College—point out that the demand for scientific evidence through something like laboratory testing is in effect asking God, the Supreme Being, to become our servant.[9] Kreeft and other philosophers argue that the question of God should be treated differently from other knowable objects in that this question regards not that which is *below* us, but that which is *above* us. As Benedict XVI put it: "The arrogance that would make God an object and impose our laboratory conditions upon him is incapable of finding him. For it already implies that we deny God by placing ourselves

---

[9]   Peter Kreeft and Ronald K. Tacelli, *Handbook of Christian Apologetics: Hundreds of Answers to Crucial Questions* (Downers Grove, IL: IVP Academic, 1994), 264.

above him."[10] The pontiff goes on to say, "The highest truths cannot be forced into the type of empirical evidence that only applies to material reality."[11]

Such a stand raises the discussion to a "higher level." The question "Does God exist?" is not like "Do neutrinos exist?" God cannot be "trapped" by some kind of ingenious experiment. If God resides on a level different from scientific issues—not measurable, quantifiable, or touchable—then empirical proofs of the experimental type are obviously out of the question to prove God's existence.

The third reason is that God cannot be treated as some kind of hypothesis the way scientists test their hypotheses. We could never call God a hypothesis that we stumbled upon or came up with in our scientific endeavors. Because hypotheses are always open to disproof and thus should be only tentatively held, God is not a hypothesis that we hold on to tentatively and provisionally until more evidence for or against it emerges.

The fourth reason is that, as Chesterton once said: "Atheism is indeed the most daring of all dogmas.... It is the assertion of a universal negative; for a man to say that there is no God in the universe is like saying that there are no insects in any of the stars."[12] Chesterton is right; it is much easier to establish that there is a black swan somewhere on earth than to prove that there are none at all. We may perhaps validly conclude that God is unknown (as agnosticism asserts), but it is very hard, if not logically impossible, to conclude that God is, in fact, absent (as atheism claims). It is impossible to close a search for God with the conclusion that there is *no* God. No search ever conclusively

[10] Joseph Ratzinger, *Jesus of Nazareth* (San Francisco: Ignatius Press, 2007), 36.
[11] Ibid., 216.
[12] G. K. Chesterton, *Varied Types* (self-pub., 2017), 24.

reveals the *absence* of its object. Absence of evidence is not evidence of absence.

The fifth reason is that it's very risky to put all our trust in our five senses. There may be much more to life than what "meets the eye." Gravity, for instance, is invisible to the human eye (only its effects are visible); the same can be said about electricity, energy, and X-rays. Sometimes only the invisible can explain what is visible. So let's not put all our eggs in the basket of science.

All the above reasons haven't prevented scientists from denying or rejecting God's existence. What about the scientists whose claims go further than what you suggest? Many of them have vehemently declared that science proves to us that there is no God at all.

The problem is that these scientists hardly ever tell us who the God is whose existence they deny or reject. They may have very different things in mind without telling us. Let's find that out first: whom or what are they rejecting?

Are these scientists rejecting the God who is taken as being a hypothesis postulated in science? God is not a hypothesis, as we said already. In that respect, the legendary French astronomer Pierre-Simon Laplace was right. When given a copy of Laplace's latest book, Napoleon Bonaparte received it with the remark, "They tell me you have written this large book on the system of the Universe, and have never even mentioned its Creator." Laplace answered bluntly, "I had no need of that hypothesis."[13]

And remember, hypotheses in science are always tentative. Instead, belief in God's existence is of a different nature. It is more like saying about a spouse, parent, or friend, "I *know* that person exists"—which is certainly not a hypothesis either. A man named

[13] Although the conversation in question did occur, the exact words Laplace used and his intended meaning are not known.

Job, in the land of Uz, said something like this about God: "I *know* that my Redeemer lives."[14] Belief in God expresses someone's faith and commitment, which cannot be tentative without undermining the very faith it means to express. In short, scientists do have good reason to reject God as a hypothesis—because that's not what God is—but rejecting God as a hypothesis is not the same as rejecting God per se.

Or perhaps the scientists you refer to in your question are rejecting the god who replaces what science has not been able to explain. This kind of god has become known as the "god of the gaps." Scientists may have a very good reason for rejecting this kind of god, who merely fills in the gaps left behind in our scientific explanations. That's not where God belongs, however, no matter how attractive this may seem to some scientists.

Even Isaac Newton fell for this timeless wish to have God keep a "divine foot" in the door when he called upon His active intervention to reform the solar system periodically from increasing irregularities and to prevent the stars from falling in on each other.[15] When Newton called on such special interventions by the Creator in the working of the universe, the German philosopher Gottfried Leibniz quipped, "God Almighty wants to wind up his watch from time to time: otherwise it would cease to move. He had not, it seems, sufficient foresight to make it a perpetual motion."[16]

Today, we know that God does not have to make these interventions in Newton's universe, because science can now explain them with the proper laws (which are God's laws anyway). The great Scholastic theologian Francisco Suárez put it this way: "God

---

[14] Job 19:25 (emphasis added).
[15] Isaac Newton, *Opticks*, 2nd ed. (1706), query 31.
[16] H. G. Alexander (ed.), *The Leibniz-Clarke Correspondence* (Manchester, UK: Manchester University Press, 1998), 11.

does not intervene directly in the natural order where secondary causes suffice to produce the intended effect."[17]

The problem with a god of the gaps is that he is a fleeting illusion, for when the frontiers of science are pushed back—and they continually are—this kind of god is pushed back with them. Seen this way, he becomes a stand-in god who is hauled in to explain effects in the world when no better cause seems to be available. The idea behind this view is that when God is doing something, then nothing else is doing it, and when something else is doing it, then God cannot be the cause of it. Instead, as the theologian Dietrich Bonhoeffer famously put it, "We are to find God in what we know, not in what we don't know."[18] So scientists have good reason to reject the notion that God is a god of the gaps.

Or perhaps the scientists you refer to in your question are rejecting the God they know from Christianity? If so, that does not give them the right to deny or reject God's existence entirely. We will see later that different religions may have different conceptions of God, but the monotheistic religions share a belief in God's existence. True, Christians are more specific on who this God is—a Triune God, one God in Three Persons—but if someone feels entitled to reject a Triune God, that does not entitle him to reject any kind of God.

Or perhaps the scientists you refer to are rejecting God as a delusion ahead of time, without any further discussion. If so, then I'd like to ask them what gives them the right to take this stand. If God is indeed a delusion, then the question of God's existence certainly makes no more sense. But before we accept that conclusion, we must ask first what makes them think this way.

[17] Francisco Suárez, *De Opere Sex Dierum*, II, c. x, n. 13.
[18] Dietrich Bonhoeffer in a 1944 letter: *Letters and Papers from Prison*, ed. Eberhard Bethge, trans. Reginald H. Fuller (New York: Touchstone, 1971).

A Catholic Scientist Proves God Exists

Someone to blame for this thought is Sigmund Freud, who believed that the adoption of religion is a reversion to childish patterns of thought in response to feelings of helplessness and guilt. According to Freud, we supposedly feel a need for security and forgiveness and so invent a source of security and forgiveness: God. Belief in God's existence is thus seen as a childish delusion, whereas denying God's existence is taken as a form of grown-up realism.

It is very doubtful, however, whether Freud really did refute belief in God's existence as an illusion. First of all, if Freud claims that basic beliefs are the rationalization of our deepest wishes, wouldn't this entail that his atheistic beliefs could also be the rationalization of his wishes? Don't we sometimes think that which we wish is true? Second, C. S. Lewis noticed a serious inconsistency: "The Freudian proves that all thoughts are merely due to complexes—except the thoughts which constitute this proof itself."[19] Third, even if belief in God were indeed wishful thinking, one could never prove that it is nothing more than wishful thinking. The God one would like to exist may actually exist, even if the fact that one wishes it may encourage suspicion.

So, we must come to the conclusion that science is not the right tool to prove God's existence. Besides, science cannot answer all our questions, and certainly not the question of God's existence. If scientists still think they have the right to air their opinions about God's existence or nonexistence, we should at least question their opinions. No matter how some scientists defend their denial or rejection of God's existence, the question remains whether their reasons to reject a certain kind of God are reason enough to reject any kind of God. Therefore, based on other sources, we know that God exists. Next we need to find out who God is.

---

[19] C. S. Lewis, *Miracles* (San Francisco: HarperOne, 2015), 55.

2

How We Can Know God Exists

Saying we *know* something requires some kind of proof. If we have no proof, we don't really know but have only an opinion or a hypothesis. In order to know that God exists, we need proof that God exists. In general, there are at least two kinds of proofs: scientific proofs and mathematical proofs.

What would a scientific proof for God's existence look like? I would think that kind of proof is vital to convince scientists.

As I said earlier, God's existence cannot be tested in the laboratory. Is that detrimental to the claim that God exists? I don't think so, because scientific proofs may not be what they pretend to be or what you seem to think they are.

In general, scientific proofs are based on what is called *inductive* reasoning, which is a way of reasoning that is not logically safe and therefore not a proof. It leads us only to probable conclusions based on a series of premises—perhaps highly probable, but never certain. It is basically a form of "generalizing induction," which starts with a series of statements about singular observations made in the lab or in the field. After adding more and more similar observations, one comes closer and closer to a generalized observation statement expressed in the conclusion.

For instance, after having seen many times that iron expands by heating, one might conclude that *all* iron does so. Of course, that conclusion is not foolproof, for it is based on a finite sample of cases mentioned in the premises against an *infinite* set of all cases expressed in the conclusion. We still might come across a case that falsifies the conclusion. So, *verification*, as a seal of proven knowledge, is not really possible through generalizing induction. This made the Nobel laureate Bertrand Russell describe science as follows: "Its method is one which is logically incapable of arriving at a complete and final demonstration."[20] This also explains why science has been wrong so often in the past. Theories accepted today may be revised or even discarded tomorrow. The history of science shows us many instances, but I will mention only four.

1. Vulcan was a planet that nineteenth-century scientists believed to exist somewhere between Mercury and the sun, but Vulcan turned out to be a phantom.

2. The expanding-Earth hypothesis stated that continental drift could be explained by the fact that the planet was gradually growing larger.

3. For a while, scientists commonly believed that the size of the universe was an unchanging constant until they learned that wasn't true.

4. For decades, it was believed that humans have forty-eight chromosomes until it turned out to be forty-six.

There are numerous other cases of revision or even rejection of scientific theories and ideas. The bottom line is this: science is of an inductive nature and cannot provide final certainty.

---

[20] Bertrand Russell, *Religion and Science* (Oxford, UK: Oxford University Press, 1954), 8.

Of course, scientists realize the shortcomings of inductive proofs in scientific research. Thus, they often favor another tool, *falsification*, which is a deductive form of reasoning and therefore logically safe. Falsification says that a hypothesis or theory is in trouble when its predictions turn out to be false. The champion of this idea, the Austrian-born physicist Karl Popper, put it this way: "Every 'good' scientific theory ... *forbids* certain things to happen."[21] This rule is based on a deductive, logically valid way of reasoning: if X is true, then Y is true; well, Y is *not* true; therefore X *must* be false.

Not surprisingly, Popper made falsifiability or refutability a requirement for scientific theories. Hypotheses and theories that do not forbid something shouldn't be accepted in science. Albert Einstein said something similar: "No amount of experimentation can ever prove me right; a single experiment can prove me wrong."[22] Falsification is basically a deductive, safe way of reasoning: if we find one black swan, a previous inductive conclusion stating that all swans are white has been deductively and conclusively falsified.

But don't get too excited too soon. There are a few drawbacks when it comes to falsification. It is always possible to reject falsification for the simple reason that falsification is based on an *observation* that is counter to what the hypothesis or theory had predicted. Perhaps this one black swan wasn't black at all, but painted. In other words, that one observation has to be certain and unquestionable on its own, which is never possible. So ultimately, even falsifying evidence depends on verification. But if final verification is beyond

---

[21] Karl Popper, *Conjectures and Refutations: The Growth of Scientific Knowledge* (London: Routledge and Keagan Paul, 1963), 33–39.

[22] This is a paraphrase from his "Induction and Deduction," document 28 in *Collected Papers of Albert Einstein*, vol. 7, *The Berlin Years: Writings, 1918–1921*, trans. Alfred Engel (Princeton: Princeton University Press, 2002).

our reach, so must falsification be. Therefore, it is always possible to reject any kind of counterevidence as spurious.

An added problem of falsification is that it's very unusual in science for a single theory to be put to the test. Almost every hypothesis or theory in science is tied to a set of conditions and presuppositions, which may not be true. Added to this is that in most sciences, observation relies heavily on complicated instruments, and this makes the impact of additional theories even more prevalent. Most falsifying observations owe their existence to some kind of measuring equipment and some theoretical background—and either one of these may be wrong or in need of some adjustment. Only the experimenter knows the many little things that could have gone wrong during an experiment.

In other words, falsification does not make a falsifiable hypothesis or theory automatically falsified on the first hit. We may have thought we had a new motto for science—"Although science cannot prove, it can disprove"—but now we have to add that science cannot even "disprove" in a rigorous way. When nature says no to our tests, it is not exactly clear what exactly it says no to.

Another serious problem for science is that the same phenomena can be explained by multiple theories at the same time. For example, the geocentrism of Ptolemy, the heliocentrism of Copernicus, and the mixed system of Tycho Brahe all enabled sailors accurately to navigate the waters, as they could all explain what was seen in the skies. Yet two of them turned out to be wrong.

Let's come to a conclusion. In science, we don't receive verification, but rather confirmation at best. Verification is, in essence, an unattainable goal for scientists—there is no possibility in science of "knowing for sure." The most scientists can claim is that the hypothesis under consideration is extremely likely after they have received many similar confirming test results for their predictions. But final certainty remains beyond their reach.

It is often said that science, if properly understood, still gives us pointers to God's existence. You seem to ignore those pointers. Didn't St. Paul say about God that "since the creation of the world his invisible nature, namely, his eternal power and deity, has been clearly perceived in the things that have been made"?[23]

Indeed, you used the right term—*pointers*. They point to God, but that does not mean they prove beyond any doubt that God exists. They are like indications or clues we find, but not proofs—at best, they prove beyond *reasonable* doubt, but they don't prove beyond any doubt or anyone's doubt. What is "reasonable" to some may not be reasonable to others. So we are still in limbo.

Yet you are right that this kind of confirming evidence based on pointers does indeed exist. That is possible even in science. Perhaps a good example is the evidence biologists give us for evolution, understood as "common descent with modification." Do all biologists accept that there is evolution? No, there are still quite a few dissenters who keep stressing that we have never seen a species change into another species. And yet there are many indications that there *must* be something like evolution; these "pointers" come from paleontology, taxonomy, anatomy, embryology, genetics, and other fields. Put all together, the evidence becomes stronger and stronger—but never indisputable beyond any doubt.

The legendary biologist Theodosius Dobzhansky summarized the situation well: "Nothing in biology makes sense except in the light of evolution."[24] But "making sense" is a rather vague form of evidence. Does this qualify as irrefutable proof? Certainly not. It's more like a proof beyond reasonable doubt. But it does not give

---

[23] Rom. 1:20.
[24] Theodosius Dobzhansky, "Nothing in Biology Makes Sense Except in the Light of Evolution," *American Biology Teacher* 35, no. 3 (March 1973): 125–129.

us the certainty scientists would like to have. What we have here is not a matter of proof but of credibility.

Proving God's existence based on all the indications we find in nature, and even in science, is of a similar nature. Such indications may provide some evidence for God's existence, but whether it is sufficient evidence remains always questionable. Many scientists will consider the evidence insufficient, as some biologists find the evidence for evolution insufficient. Empirical proof "beyond any doubt" is hard to come by—even more so when it comes to God's existence. The best we can say, paraphrasing Dobzhansky's line, is this: nothing on earth makes sense except in the light of God's existence.

Besides, we can't expect science to give us empirical proof of God's existence—that's not its territory and expertise. Recall that science cannot even declare its own scientific findings as final and proven. What some call "proven" scientific knowledge is proven only until a new set of empirical data "disproves" it. So "empirical proof" is not cast in iron; whatever is true today may not be true tomorrow. Science is always a work in progress. Asking for final proof is asking for more than science can ever deliver, especially when it comes to proofs of God's existence.

You seem to be telling me that science could never prove God's existence, at least not with proofs in the strict, technical sense. Is there a way for us to know with certainty that God exists?

There must be other ways of "knowing" than the ways science uses. I mentioned already the way we "know" things in mathematics. Mathematical proofs are the kind of proofs scientists can only dream of in science. Earlier I said that there are inductive and deductive ways of reasoning. Well, the deductive way is typical for mathematics.

Proofs in mathematics may seem to be safe and foolproof, but that's so only in a rather "artificial" way. Mathematical knowledge may seem the most secure form of knowledge, but it is not about anything "real." Albert Einstein said it right: "As far as the laws of mathematics refer to reality, they are not certain; and as far as they are certain, they do not refer to reality."[25] When we add one to one in math, the result is necessarily two $(1 + 1 = 2)$; but when we add one drop of water to another drop of water, the result is not two drops but one drop $(1 + 1 = 1)$; and when we add one organism to another, we may end up with three or more of them $(1 + 1 \geq 3)$.

Besides, even in mathematics, we cannot prove anything from nothing, for at least something is needed—axioms—to start the process. Based on a set of axioms in the premises, we can derive a conclusion with final certainty. Based on such a set of axioms, it can be concluded, for example, that the sum of the three angles of a triangle is 180 degrees. The certainty of the conclusion is based on the certainty of the axioms. That's quite a trivial way of proving things. Yet it's still more certainty than we can ever reach in science.

How could the certainty we find in mathematics help us when it comes to proving God's existence? I don't think mathematics would be our tool.

The "secret" of proving God's existence is not to be found in mathematics but in logic. It can be found in so-called *universal* statements. They are probably best understood as something like safe starting points or assumptions. They are not the result of empirical induction (as in science); neither are they the outcome of deductive proof (as in math). Instead, they are said to be self-evident—meaning "not in need of any further evidence." They can provide certainty to the premises, which then extends to the certainty of the conclusion.

To get this right, do not confuse universal statements with the generalizing statements we find in science— statements such as

---

[25] Address to Prussian Academy of Sciences, 1921.

"All iron expands with heat." Generalizing statements are confirmed by testing more and more instances of iron under various conditions; such a case is an example of inductive generalization. In contrast, universal principles, such as "All expanding of iron has a cause," are true, independently of any particular cases; their truth does not increase by testing more and more instances, because their truth does not depend on any confirmation, let alone inductive generalization. It is a universal principle that is true independently of any observation.

The principle that like causes have like effects comes before science can even get started. It's a universal principle or statement that is self-evident, admits no exception, and cannot be denied by anyone in his right mind.

In a sense, it is similar to the situation in mathematics. Michael Augros uses a good analogy from mathematics: "I have not inspected every instance of the number six, yet I am convinced ... that the general statement 'Every six is even' admits no exception. I am not worried that someone in Australia has an odd six in his pocket."[26] In a similar way, we don't have to worry that someone comes up with a case that has no cause. We tell people who claim they have found a case without any cause to keep searching for a cause. This universal statement cannot even be falsified.

How could the distinction you introduced here between general statements and universal principles be helpful in your effort to find a real, conclusive proof of God's existence?

To begin with, the good news is that we are capable of grasping the truth of universal principles without the assistance of science. The truth of such universal principles is accessible to everyone — they

---

[26] Michael Augros, *Who Designed the Designer? A Rediscovered Path to God's Existence* (San Francisco: Ignatius Press, 2015), 18.

are self-evident—whereas the truths of science must be discovered, or they must be received from others who discovered them before us.

So the question is now: What would make a good starting point for proofs of God's existence? Well, universal statements are what we need here. I would say that everyone must agree with the following universal statement: everything that comes into being must have a cause. This is a statement that no one can deny, even though we have not inspected every instance it covers. No people in their right minds are able to deny this—it's self-evident.

This universal statement explains, among other things, that you cannot give existence to yourself or receive it from yourself. In short, you cannot be your own parent; children need parents—which can start a long chain or sequence of generations. In theory, this sequence of generations can stretch backward in time as well as forward in time. It may have a beginning and an end, but it may also stretch infinitely far back or forward without any beginning or end—it doesn't really make a difference.

We have a similar situation with the chain of events running in the history of the universe. This chain may be finite, with a beginning and end—as is the case in the Big Bang scenario—or it might be infinite, having no beginning or end—which was thought to be the case by many physicists at the beginning of the previous century, including Albert Einstein. Either way, the chain has a sequence of events with causes, following each other in time. It is not essential for our discussion whether the sequence is finite or infinite. All that matters is the universal statement that says, "Everything that comes into being or existence must have a cause."

But there is something even more essential about this sequence, whether finite or infinite. Though all causes in a chain keep each other in tow, there is a more fundamental question: What is it that keeps the entire chain going? In other words, what is the cause of the entire chain itself—given the fact that everything that comes into existence must have a cause? This question is not about causes working inside the chain but about a cause operating from outside the chain.

I wonder, though, why would we need "an outside cause of the inside causes"? That sounds to me like endlessly piling up explanations on previous explanations, which is basically a form of infinite regress with no end in sight.

Let me explain first why having a sequence of causes is not enough to explain the entire chain. Let's use again the example of a chain of parents and their children. As I said before, it doesn't really matter whether the chain is finite or infinite. If the chain is finite, there is one parent to start the chain—the first parent, so to speak. But even if the chain is infinite—that is, going back in time without a beginning—there is no parent to start the chain, but there is always a parent going further and further back in the chain of parents. In either case, we might think there is always a parent to explain what happens inside the chain. Really? It may seem so, but that one parent doesn't really explain the entire chain.

Why not? Here is the crucial point. No parent, not even a "first parent," can explain everything about parents and children, for the question remains: What is it that enables parents to generate children at all? Where does that generational power come from? Moreover, a sequence of events—either finite or infinite—is possible only if there is something such as time, for the parent must come before the child in time. Where does time come from then? That question calls for another cause—the cause of time—to explain where time itself comes from. In other words, the existence of the entire chain calls for causes or explanations at a higher level. Causes outside of the chain make causes inside the chain possible.

We must conclude from this that the causes inside a chain cannot fully explain the chain itself. In other words, the chain of causes needs to be hooked onto something else, so it won't hang from nowhere, just floating in the air. Put differently, what is it that keeps this sequence or chain of causes going? There must be an ultimate cause—at a higher level—that explains all other causes. That ultimate cause is usually called the First Cause, and it is needed to make other causes, the secondary causes, possible.

Perhaps the following analogy will help us understand this idea better. Think of an I-beam with a hook on it from which a chain is to be hung, suggests the philosopher Michael Augros:

> If there is nothing for that whole chain to hang from, it will not hang, and nothing can be hung from it. There is nothing about those links in themselves that makes them want to hang in space.... There must also be something from which things hang and which is not itself hanging from anything.[27]

The I-beam is like a "first cause" that explains the secondary causes that keep the chain together.

A more intricate analogy is the following. Think of a book that is sitting on a bookshelf.[28] The book has no capacity of its own to be five feet from the floor; it will be there only if something else, such as a bookcase, holds it up. But the bookcase, in turn, has no power of its own to hold the book there. It, too, would fall to the earth unless the floor held it aloft. And the floor, for that matter, can hold up the bookcase only because it is itself being held up by the house's foundation, and that foundation, in turn, by the earth, and the earth by the structure of the universe. All these "intermediaries" keep one another in tow.

[27] Augros, *Who Designed the Designer?*, 35.
[28] I borrowed this example from Edward Feser, *The Five Proofs of God's Existence*, 21.

A Catholic Scientist Proves God Exists

None of these intermediate things, however, could hold up anything at all without something to hold them all up without having to be held up itself. That ultimate "something" is best called the First Cause of all secondary causes. Without it, nothing in this series of causes would really be explained, let alone exist. This First Cause has the power to produce its effects without being caused by something else. It has *inherent* causal power, while the secondary causes have only *derived* causal power.

By speaking of an "ultimate cause," we avoid infinite regress, which was mentioned in an earlier question. This maneuver, however, is not just a matter of making sure that "the buck stops somewhere." It's not an arbitrary move, and it's not unreasonable at all. I will show later in this book the rationality behind it. For now, let's assume that the First Cause is the ultimate cause of all there is in this world.

It is very reasonable to identify this First Cause with God. But let's not jump to conclusions yet. St. Thomas Aquinas, who uses the term *First Cause* frequently in his works, is very careful when he says at the end of each proof of God's existence, "And this all think of as God." In his five proofs of God's existence, Aquinas does not use *god* as a proper name but as a common noun. So each one of his proofs concludes only that there is "*a* god."

No matter how we look at this, we have here a compelling, deductive way of proving that God, or at least "a god," exists as a First Cause. Obviously, it needs more "body" than we gave it here, but we will work on that in the next chapters. Let's call the proof given above a metaphysical proof to distinguish it from any physical proof or even any mathematical proof. These metaphysical proofs are deductive and have as much, if not more, power than scientific proofs, which are of an inductive nature. Metaphysical proofs can prove what scientific proofs cannot prove with certainty—which is something they seem to have in common with mathematical proofs. Let's find out in the next chapters what their real power is.

3

Proofs of God's Existence

Proofs of God's existence cannot come from science, as we found out, for science deals only with material entities—so God is beyond its reach. Besides, science is an inductive enterprise, so scientific proofs are never final proofs beyond any doubt. Therefore, any proof of God's existence must be deductive in nature, which calls for a metaphysical proof.

There have been several of these proofs. We will discuss only the five best-known ones.

The Argument from Existence

Do electrons and viruses exist? The answer from science is yes. Do UFOs and cosmic rays exist? The answer is up for discussion. Do mermaids and unicorns exist? The answer is a definite no. The difference between these answers is essentially a matter of *existence*.

"To be, or not to be, that is the question"—not only in Hamlet but also in our discussion here. In metaphysical terms, everything has two principles that explain its being (*esse*): essence (*essentia*) and existence (*existentia*). In all finite beings, these principles are both required in order for the individual thing to be. Essence may be described as the "what" of a thing, that which is known about it by our forming a concept so as to understand what it is. It is a universal principle that makes material individuals, for instance, to be of the same kind—being material, that is.

But it is obvious upon reflection—actually self-evident—that "*what* a thing is" and "*that* a thing is" are completely different. Having an essence doesn't guarantee a thing's existence. Anyone can understand what a unicorn or a mermaid is but may yet not know whether these things exist. Fairy tales play on this distinction. Therefore, a thing's existence—*that* it is—is obviously and evidently different from its essence—*what* it is.

Apparently, understanding an essence does not necessarily include understanding its existence. There is a real distinction between essence and existence. If existence were part of the "what" of a thing—that is, part of its essence—then that thing would necessarily exist, including mermaids and unicorns.

But we know that this is not the case. The act of existing is something received. All existing things have a received existence. Each one of them had to come into existence, independent of its essence.

What you say here sounds rather trivial to me: there are things that exist and things that do not exist. That doesn't tell us much. Applied to God: God either exists or doesn't exist. We knew that already. That's not a proof of God's existence, is it?

No, it's just a starter. When things come into existence, they need a *cause* that makes them exist. X-rays, for instance, exist because radioactive decay makes them exist, and this decay is caused by radioactive isotopes, and radioactive isotopes are caused by unstable nuclei, and so on. All of these causes bring their effects into existence. But all of them combined do not really explain the existence of X-rays. They lead to infinite regress, calling for more and more causes further and further down the ladder of explanation. They can never finish their job—they leave the ladder resting on nothing or hanging from nowhere, so to speak.

To solve this problem, we need to start our argument for God's existence with a universal statement that is *self-evident*. That statement would look like this: nothing can make itself exist. Or put differently: everything needs a cause for its existence. Or perhaps even better: everything that comes into existence has a cause.

There is no need to test this principle or to corroborate it with repeated confirmations, for it is not of the kind that science achieves with its general statements. It is a principle that is universal and self-evident. We use it and need it before science can even start searching for causes and explanations of things that exist. It makes no sense to search for causes if there are no causes that make things exist.

It seems to me that I can certainly just deny the universal statement you seem to consider as a self-evident statement. I hardly ever hear people, let alone scientists, mention this principle.

You are right; it's hardly ever mentioned. I would say it is so self-evident that it can easily "hide" as an assumption that remains unspoken or even unnoticed. There is nothing strange about that. All people, and scientists in particular, hold assumptions that are taken so completely for granted that they are never pointed out. We mentioned several in the previous chapters.

If you do deny this principle, however, you basically opt for a new assumption. The most common assumption among many people, especially scientists, is that *matter* can cause itself and explain itself. What's wrong with that assumption? Unlike the previous one, this one is not self-evident and is based on some kind of magic. The idea that matter can cause itself and explain its own existence has rightly been caricaturized by Peter Kreeft as a magical "pop theory" that claims that things pop into existence without any cause.[29] Nothing can just pop into existence; it must have a cause, because it does not and cannot have the power to make itself exist. Nothing can generate itself, just as children cannot generate themselves.

Is that really self-evident? How do we know this is true? Well, for something to generate itself, or produce itself, it would have to exist before it came into existence—which is absurd. Therefore, whatever came into existence must have a cause outside itself that caused its existence. This is basically a deductive form of reasoning—sometimes called *reductio ad absurdum*.

In this sort of argument, we begin by assuming the opposite of the claim we want to prove, and we show that from this claim something absurd follows. This shows that the opposite of the claim we want to argue *for* is false, and so the claim we want to argue *against* is true. In our case, claiming that something can make itself exist amounts to a kind of circular causation. That would be absurd and must therefore be rejected. You can't give yourself existence or receive existence from yourself. As Michael Augros puts it pithily, "You can't be your own father."[30]

Not only is it impossible for anything that has come into existence to *cause* itself, but it is also impossible for anything that has come into existence to *explain* itself. In other words, nothing that has come into existence is self-explanatory—not the universe, not man, not even matter. Notice how the focus of the argument has shifted here slightly from "everything that comes into existence has a *cause*" to "everything that comes into existence has an *explanation*." So, instead of searching for causes, we might as well search for explanations.[31]

---

[29] Peter Kreeft, *The Creed: Fundamentals of Christian Belief* (San Francisco: Ignatius Press, 1988), 27.

[30] Augros, *Who Designed the Designer?*, 33.

[31] For the record, all causes are reasons or explanations in the sense of making their effects intelligible, but not all reasons are causes.

This latter search has been propelled most notably by an intriguing question first posed by the German philosopher Gottfried Leibniz some three centuries ago: "Why is there something rather than nothing?"[32] His point was that there must be an explanation (or a reason) for why there is something rather than nothing.[33] Without any further explanation, it is a mere riddle why there *is* something—that is, why something exists. It cannot explain its own existence.

Asking why there is something rather than nothing seems to me a pretty senseless question. My answer to this question of why there is something is simply that this something apparently exists. Why is it there? Because it is there! We don't need any further explanation.

I think we do need more of an explanation. Things cannot cause themselves, so we found out, but neither can they explain themselves. To say, as scientists like to do, that there is something because of something else—its cause or explanation, that is—only shifts the question from one thing to another. We then end up with a series of explanations—one explanation on another explanation, with no end in sight. But we never reach a *sufficient* explanation or reason. Leibniz concluded from this that a sufficient reason, which needs no further reason or explanation, must be outside this chain of explanations and must be found in a necessary being bearing the reason for its existence within itself.[34]

In other words, there must be an explanation that is not in need of any further explanation. Nothing less than an infinite,

---

[32] G. W. Leibniz, *The Principles of Nature and Grace, Based on Reason* (1714).

[33] An unfortunate implication of this wording is the possibility that there could have been nothing at all, including no God. It would have been better to limit it to things that have come into existence.

[34] Leibniz, *The Principles of Nature.*

self-explanatory, necessary being could possibly terminate the regress of explanation. This being is "necessary," as a mathematical truth such as "eleven is a prime number" is necessary and could not be otherwise. Only a necessary, self-explanatory being can explain with "sufficient reason" why this world exists. No other explanation qualifies as sufficient. So what we have here is basically a proof of God's existence.

You seem to imply that only God qualifies as a final, ultimate explanation of this world. But is that true? Aren't there other candidates for this supposedly unique position?

It should not come as a surprise that other "candidates" have indeed been postulated to bear the reason for their existence within themselves. One popular candidate has been the universe itself, in particular its laws of nature, as the ultimate explanation of all that exists. Stephen Hawking, for one, talked about the Big Bang in terms of something self-explanatory: "Because there is a law such as gravity, the Universe can and will create itself from nothing. Spontaneous creation is the reason there is something rather than nothing."[35] How could this possibly be true? How could the universe be the explanation for its own existence?

First of all, the law of gravity cannot do the trick, for, before the universe existed, we would have to posit laws of physics—which are ultimately the set of laws that govern the existing universe. Since the laws of nature presuppose the very existence of the universe, they cannot be used to explain the existence of the universe. So Hawking is actually saying that laws that have meaning only in the context of an existing universe can generate that universe, including the laws of nature, all by themselves before either

[35] "Stephen Hawking and Leonard Mlodinow, The Grand Design," *Times Eureka*, no. 12 (September 2010): 25.

exists—which not only is a form of circular reasoning but also makes for an absurd conclusion.

Second, the laws of nature could have been other than they are—they could exist or not exist; they could be this or that. It is easy to think, for instance, about possible worlds in which the laws of nature are radically different from those that are operational here and now in our universe. The universe could even have been "law-less." Physicist Paul C. Davies is right:

> There are endless ways in which the universe might have been totally chaotic. It might have had no laws at all, or merely an incoherent jumble of laws that caused matter to behave in disorderly or unstable ways.[36]

Instead, we find in this universe a very specific "law and order." In other words, the universe doesn't have to behave the way it does. Only a self-explanatory, necessary being—God—can explain why the laws of nature are the way they are.

Third, the question of why there are laws of nature at all—rather than nothing—arises. Paul Davies comments:

> Over the years I have often asked my physicist colleagues why the laws of physics are what they are?... The favorite reply is, "There is no reason they are what they are—they just are."[37]

That's a moot reply, for these laws could have been other than they are, so they are not self-explanatory. If there is no inherent necessity for the universe to exist, however, then the universe, including

---

[36] Paul Davies, *The Mind of God: The Scientific Basis for a Rational World* (New York: Simon and Schuster, 1992), 195.

[37] Paul Davies, "Taking Science on Faith," *New York Times*, November 24, 2007.

everything in it, is not self-explanatory and therefore must have an explanation outside itself. Obviously, it cannot be explained by something finite, not necessary, and not self-explanatory—for that would lead to infinite regress—so it can be explained only by an unconditioned, infinite, necessary being.

Fourth, the universe itself cannot be a necessary being. If it had to exist necessarily, the universe could not be different from what it is. But what is worse, everything about it could be deduced by pure thought without any need for further observation and exploration. That implication is obviously absurd and would make science completely redundant. That's another case of a *reductio ad absurdum*.

Obviously, we are dealing here with a philosophical issue, not a scientific one. Science can never explain what science itself must take for granted. That is quite obvious when you think about it. Interestingly enough, even someone like Stephen Hawking had a clear-headed moment when he said:

> What is it that breathes fire into the equations and makes a Universe for them to describe? The usual scientific approach of science of constructing a mathematical model cannot answer the questions of why there should be a Universe for the model to describe. Why does the Universe go to all the bother of existing?[38]

He was basically facing the question, head-on, of why there is something (rather than nothing). The physicist Paul Davies goes as far as speaking in terms of "the mystery of ... why there is a universe at all."[39]

---

[38] Hawking, *A Brief History of Time* (New York: Bantam Books, 1998), 174.

[39] P. C. W. Davies, *God and the New Physics* (New York: Simon and Schuster, 1983), 42.

We must come to the conclusion again that we don't need science—that we can't even use science to prove God's existence. A proof like that had to come from somewhere else. Science has no answer for everything and doesn't know all there is! It needs something outside itself—an ultimate cause or explanation. The argument from existence is one of the answers to the question of God's existence.

The Argument from Contingency

In the previous argument, we made a distinction between the essence of something and its existence. Without this difference, the things we know from experience would exist in a necessary way. But they don't. Water could exist or not exist. Quarks could exist or not exist. Genes could exist or not exist. If any of these were bound to exist, they would be necessary.

If that were so, then there would be no need for science, there would be no need for formulating hypotheses to find out whether certain things exist, and there would be no need for doing research—everything would be determined ahead of time. But that's not the way it is. Instead of being necessary, the things science deals with are *contingent*—which means they could be different; they could either exist or not exist.

So their existence must be caused or explained by something else distinct from it. That's where the argument from contingency comes in. This argument uses the universal principle of contingency as a starting point: all things that come into existence could have been *different* and could easily *not* have existed.

This starting point is self-evident again. It does not depend on a large number of supporting cases, for its truth can be seen without the assistance of science—it is, in fact, prior to science and makes scientists, for instance, search for causes and explanations of why things exist and how they can differ from each other. We don't need science to prove this principle; on the contrary, science needs this principle before it can even get started.

The point of departure is that our universe or anything in it need not be the way it is—and need not even exist. There is no inherent necessity for the universe or for anything in it to exist and to be the way it is. All beings in the universe are *contingent* beings that could easily not have existed, as the reason for their existence cannot be found within themselves—they are not self-explanatory, nor can they make themselves exist. Every contingent entity must have a cause or explanation for it to exist and to be the way it is. Of course, we could find such causes or explanations in other contingent entities in the universe, but that would lead to infinite regress—for example, going from molecules to atoms to atomic particles to subatomic particles, and further and further down the ladder of explanations.

Ultimately, such a chain of causes or explanations fails to succeed. To avoid this infinite regress, it can easily be argued that, given the fact that all existent things in the universe depend on other things for their existence, there must exist at least one thing that is not contingent and therefore not in need of any other cause. This must be a *Necessary* Being, whose essence is identical to its existence. This Necessary Being *is* Existence Itself, and it is the only possible explanation and cause of all contingent beings. This makes for another proof of God's existence.

Don't scientists have explanations, though, as to why the universe exists—with the Big Bang theory, for instance, being one of them? Those explanations may not be complete, or they may not even be correct, but they are explanations. What more would we need?

The question remains, though, as to whether scientific explanations are "sufficient reason" for the existence of the universe and everything in it. The problem here is that no scientific explanation can explain why something contingent exists, for science deals, per definition, only with things that are contingent and exist already. But why do they, in fact, exist? The reason for their existence cannot be found in themselves.

This problem has been mentioned by many people. Here are just two of them, the first one from the camp of philosophy and the second from the scientific camp. David Bentley Hart puts it this way: "Physical reality cannot account for its own existence for the simple reason that nature—the physical—is that which by definition already exists."[40] Or, in the words of particle physicist Stephen Barr, "Anything whose existence is contingent (i.e., which could exist or not exist) cannot be the explanation of its own existence."[41]

If there is no inherent necessity for the universe to exist, however, then the universe, including everything in it, is not self-explanatory and therefore must find an explanation outside itself. All *contingent* beings could easily not have existed, so they cannot make themselves exist. Neither can they explain themselves, so they are not self-explanatory. Obviously, they cannot be grounded in something else that is also finite, contingent, and not self-explanatory—for that would lead to infinite regress, which is, in essence, no explanation at all. Instead, they can only derive their existence from an unconditioned, infinite, and necessary ground;

[40] David Bentley Hart, *The Experience of God: Being, Consciousness, Bliss* (New Haven, CT: Yale University Press, 2014), 96.

[41] Stephen M. Barr, "Modern Physics, the Beginning, and Creation: An Interview with Physicist Dr. Stephen Barr," *Ignatius Insight*, September 26, 2006, https://www.catholiceducation.org/en/science/faith-and-science/modern-physics-the-beginning-and-creation-an-interview-with-physicist-dr-stephen-barr.html.

this is called a "Necessary Being," which all contingent beings depend on.

Without this Necessary Being, there couldn't be any contingent beings, for we would never be able to explain why contingent entities exist. No contingent things or series of contingent things can explain why and how contingent things can exist to begin with, for they are all contingent. A Necessary Being, which is not contingent itself, is needed as the cause and explanation of everything other than itself. Without a necessary being, the entire world of contingent beings would have no basis at all; it would collapse on the spot.

I still see at least two inconsistencies in the argument from contingency. First, the argument maintains that a Necessary Being does not need a cause for its existence, while claiming all other beings do. Second, the argument maintains that everything has an explanation, but then it makes an exception in the case of a Necessary Being.

Your first inconsistency is not an inconsistency at all. As we made clear from the very beginning, all things that *come into existence* need a cause of their existence. But a Necessary Being, by definition, does not come into existence—it is pure Existence Itself, and is the very source of all existence, having inherent causal power. It *makes* contingent beings exist on their own, giving them derived causal power. A Necessary Being makes contingent beings possible—otherwise they could not exist at all. Therefore a Necessary Being does not need a cause for itself to exist. It is pure existence. So there is no inconsistency here at all.

Your second inconsistency—that everything has an explanation, with the exception of a Necessary Being—is not an inconsistency either, for the existence of a Necessary Being does not lack an explanation. The explanation lies in its own nature as being that which is pure Existence Itself, whereas the existence of the universe cannot be explained in terms of its own nature. As Edward Feser puts it, "The difference between God and the world then is not that one has an explanation and the other lacks it, but rather that one is self-explanatory while the other is not."[42] I would say all of this makes perfect sense.

Sometimes you replace, perhaps unknowingly, the term "Necessary Being" with the term "God." What is it that tells us this Necessary Being is identical to what most people call God? In theory, it could be anything else, even mere matter. Many atheists, for example, think there is some underlying basic force—something like matter or energy—that simply exists, of which nothing else is the cause and which is the cause of everything we see. For them, this Necessary Being must be Matter (with a capital M perhaps), rather than God.

I agree, matter seems to be an attractive candidate for this "Necessary Being." Most people have some intuitive idea of what *matter* stands for—they probably think of it as "stuff." That's exactly where its appeal lies. Its strength is that it centers on one of the most noticeable elements in the world around us. This "stuff" can usually be seen, touched, heard, tasted, and smelled by our five senses—it seems to be all there is. Studying this kind of "stuff" has been vital for our survival. The fact that the world we perceive through our senses and all the things we can picture are made of matter gives great credibility to the idea that the world of real existences is entirely material and that nothing nonmaterial really exists, because we do not have senses for it.

The belief that *everything* is matter and that matter is *all* there is usually called *materialism*. It claims that the only thing that exists is matter, that all things are composed of matter, that everything is the mere result of material interactions, and that anything that is not

[42] Feser, *Five Proofs*, 168.

matter cannot exist. This claim makes materialism not only very persuasive but also very pervasive, suggesting that matter must be the fundamental substance in nature. So, in this view, matter must also be the one and only necessary entity our world depends on.

Can matter really be the "Necessary Being" we are looking for as the basis of all that exists? It is hard to believe so. First of all, *matter* is a very vague term that can be given almost any meaning. The concept of matter changes constantly, depending on who uses the term. It has also changed in response to new scientific discoveries. In physics, for instance, *matter* is losing its prominent position, because the term *mass* is well-defined, but *matter* is not. The nuclear physicist Werner Heisenberg went as far as to say, "The ontology of materialism rested upon the illusion that ... the direct 'actuality' of the world around us can be extrapolated into the atomic range. This extrapolation, however, is impossible.... Atoms are not things."[43]

Second, explaining matter in purely material terms as required by materialism is no explanation at all, but is more like a circular argument. Material things cannot account for their own existence, so we found out, for the simple reason that matter is that which, by definition, already exists. So, we do need an explanation for the existence of matter, as matter itself cannot provide that explanation. It is exactly here that a serious question should come up: Why is there something like matter rather than nothing?

Those who say that questions like these are pointless are adopting something like a "black-box hypothesis"—no more questions asked! But that hypothesis is hard to accept. To say that things simply exist and will continue to exist all on their own ignores the question of what gives them this remarkable

---

[43] Werner Heisenberg, *Physics and Philosophy* (London: Ruskin House, 1959), 128.

feature. We always tend to find explanations for why things are the way they are—and even if we don't, we tend to have reason to think there must be *some* explanation. Claiming there simply is *no* explanation at all is irrational, even absurd. Besides, I am giving you one in this book.

Third, materialism claims that nothing can exist without the materials out of which it is made. The physicist Carl Sagan worded this view with great flair when he said that the material world is "all that is, or ever was, or ever will be."[44] Matter, however, is anything that has mass and takes up space, so it is subject to motion and change. Because matter undergoes change, this change requires a cause or explanation. Matter is contingent, not necessary. If it were necessary, then everything about matter could be deduced by pure thought, and this we declared already as absurd. That's the simple reason why matter is contingent and cannot be a Necessary Being, for a Necessary Being is, per definition, immutable, eternal, and nonmaterial, according to Thomas Aquinas.[45] Michael Augros brings this argument to a close by saying, "Matter itself is a *product*, receiving its very existence from the action of something before it."[46]

From this follows that materialism is a metaphysical position; it cannot even be a conclusion of the empirical sciences. The legendary biologist J. B. S. Haldane worded it this way, "If materialism is true, it seems to me that we cannot know that it is true. If my opinions are the result of chemical processes going on in my brain, they are determined by chemistry, not the laws of logic."[47]

---

[44] Carl Sagan, *Cosmos* (New York: Random House, 1980), 4.
[45] *Summa Theologica*, I, Q. 9, art. 1.
[46] Augros, *Who Designed the Designer?*, 63.
[47] J. B. S. Haldane, *The Inequality of Man* (New York: Penguin Books, 1932), 162.

A Catholic Scientist Proves God Exists

In the proofs of God's existence given so far, God keeps coming out as a very abstract entity — definitely not a material entity. It seems to be a God of reason rather than a God of faith.

You are right on that point. Reasoning based on logic doesn't give much flesh to its conclusions. That's the price we pay for the logical power of deduction, which leads to rather abstract conclusions. So the best way to characterize a Necessary Being in the argument from contingency is the following: to be the cause of everything other than itself, it must be at least eternal, nonmaterial, immutable, all-knowing, all-powerful, and perfect. Later on, we will delve deeper into these attributes. All I will say for now is that the God of faith is more than, but certainly not less than, the God of reason.

4

The Argument from Motion

This argument is probably the best known, but also the most mis-understood, proof of God's existence. It is strongly associated with St. Thomas Aquinas, who discussed it with four other proofs of God's existence. He called the five of them "Five Ways" (*Quinque viae*) in his famous *Summa Theologica*.[48]

The argument from motion uses the universal principle of *motion* as a starting point: nothing that moves can move unless it is moved by something else. As with other proofs of God's existence, St. Thomas argues from this that there has to be a "First Mover," which he identifies as God, who is the explanation of all other movers.

I respectfully disagree with St. Thomas. His argument from motion reminds me of a primitive worldview and an outdated form of physics. And what is worse, to think of God in terms of pulling and pushing is archaic and crude, unworthy of God.

I admit that, seen through modern eyes, this argument can easily be misunderstood as an outdated outlook. When hearing or reading

[48] To learn more about Thomas Aquinas, see the Thomistic Institute's course Aquinas 101, which can be found at https://aquinas101. thomisticinstitute.org/.

Aquinas's argument, we tend to think of billiard balls that impart motions to each other in a series of collisions, which requires a cue ball that sets all the other balls in motion. But it is very doubtful whether St. Thomas Aquinas had that in mind at all.

As a matter of fact, when St. Thomas says that A "moves" B, he is saying that A "causes" B and thus "explains" B. There is no movement through space here. St. Thomas is rather thinking in terms of cause and effect. So, instead of a First Mover, we could as well speak of a First Cause, which transforms all other movers into secondary causes.[49] It is the First Cause that causes the existence of secondary causes and makes them possible. Without a First Cause, none of the secondary causes could exist, let alone become causes of their own. Sometimes St. Thomas refers to the First Cause as God: "God is in all things the cause of being."[50] Without God, they would not be and could not exist.

Aquinas's argument is far from crude; in fact, it is very sophisticated. He makes a distinction between causes that have an explanatory relationship to one another, and causes that have a temporal relationship to one another.[51] It is a vital distinction in Aquinas's thinking. To explain this difference, I will use again the example of a book sitting on a bookshelf.

Captured in a *temporal* sequence — let's call it a *linear* sequence — we could say that the book was written by an author, and the author was the son or daughter of two parents, who were each, in turn, generated by other parents, and so on — with each step in the sequence going further back in time. Although there is a first parent in this example, that is not necessarily the case, so we

---

[49] This is explained in more detail in Stephen Barr, *Modern Physics and Ancient Faith* (Notre Dame, IN: Notre Dame University Press, 2003), 258–265. See also Feser, *Five Proofs*, 63.

[50] *Summa Contra Gentiles*, bk. 2, 46.

[51] *Summa Theologica*, I, Q. 46, art. 2.

found out. In theory, the chain could go back in time infinitely. That would be an acceptable form of infinite regress. But it is not essential for a temporal, linear sequence.

An *explanatory* sequence, on the other hand, would be very different. Let's call it a "hierarchical" sequence. This sequence does not follow a timeline, but instead goes to a deeper level of more fundamental causes or explanations. Whereas a linear series of causes can, in principle, extend backward to infinity—for the causal power of each entity need not be traced back to a first entity—a hierarchical series of causes cannot go on into infinity, for the causal power of each entity is completely *derived* and can be derived only from an entity with *inherent* causal power.

Let's return to our first example of the book sitting on a bookshelf, when it was an example of a hierarchical sequence. The book will be there only if the bookcase holds it up. The bookcase would fall to the earth unless the floor held it aloft. And the floor is itself being held up by the house's foundation, and that foundation, in turn, by the earth, and the earth by the structure of the universe. All these "intermediaries" keep each other in tow. None of these things, however, could hold up anything at all unless there were something to hold them up without having to be held up itself.

That's where the "First Mover" comes in, if understood correctly. That ultimate something is the First Cause. This First Cause has the power to produce its effects without being caused by something else. It has *inherent* causal power, whereas the secondary causes have only *derived* causal power—and that's why infinite regress of secondary causes must come to a halt.

I find the word *first* very confusing, no matter whether it is used in *First Mover* or *First Cause*. It suggests that we are dealing here with the very beginning of a series of causes or explanations. But that does not seem to be the case, based on what you are trying to say.

Right. Don't take the First Cause as the first one in a temporal, linear sequence of causes. Instead think in terms of an explanatory, hierarchical sequence of causes. Therefore, it has to be stressed that *First* should not be taken as *first* in the sense of being before the next cause in time in a temporal sequence, but rather *first* in the sense of being the source of all secondary causes—a power from which all other causes derive their causal powers. In other words, the First Cause is not temporally prior but causally prior to secondary causes. That's where a hierarchical sequence is essentially different from a temporal sequence.

To use an analogy, the term *First Cause* is a bit like the expression *first prize*.[52] This prize is usually not awarded first in time, but after the third and second prizes. It ranks before all other prizes in value or significance. Similarly, the First Cause is about causal priority, not temporal priority. Therefore, there is no need to go back in time to determine whether there is a First Cause, for the First Cause is not a spatiotemporal entity that is the starting point of a linear series of motions. It is not a thing stuck in the past, but a thing existing right here and now.

This is the First Cause, on which everything in this universe depends, according to the argument from motion. All other causes are contingent—that is, not necessary and not self-explanatory. In a temporal sequence, a first parent may long be gone, but the sequence keeps going. But in a hierarchical sequence, a First Cause is permanently needed for the sequence to continue. Without a First Cause, the hierarchical sequence would collapse.

On the other hand, even in a temporal sequence of contingent things, with each one caused by another one in the series, there is still the question of why the series exists at all, for the series is just as contingent as its individual contingent elements. That's

---

[52] See Augros, *Who Designed the Designer?*, 32.

why we always need, even in a linear sequence, a necessary, self-explanatory, uncaused First Cause to make the entire chain of secondary causes possible—whether we are dealing with a linear or a hierarchical sequence.

Thus, the complete argument from motion has the following structure:

- Contingent beings must have a deeper cause or explanation.
- Such causes cannot run in an infinite regress.
- So there must be a First Cause.
- The First Cause must be a necessary being that doesn't need a cause for itself.

You keep saying that infinite regress cannot explain anything. What's your problem with infinite regress? It is very much accepted in mathematics. For instance, negative numbers go on into infinity just as positive numbers do.

Yes, in mathematics infinite regress is quite acceptable. But mathematics is a formal enterprise, not an empirical one. It's not about causes. The number zero, for instance, does not literally cause the numbers minus one or one to be previous or next in the series of numbers. But the situation is quite different when we talk about causes in a philosophical or metaphysical context. Why is that?

First of all, a series of causes may very well be infinite when we are dealing with a *temporal*, linear sequence. Take Aquinas's example of a series of fathers and sons.[53] Each son's existence is caused by his father, going back in time. Theoretically, this sequence can even go back infinitely far in time. That would be an acceptable case of infinite regress, since the existence of each son

---

[53] *Summa Theologica*, I, Q.46, art. 2, ad. 7.

in the linear sequence is caused by his father, but not necessarily by previous fathers.

Second, things are quite different when we are dealing with an *explanatory*, hierarchical sequence instead. The question is: What keeps each son in existence at each moment? Not the father, of course, for he may have died already. Without some ultimate conserving cause that keeps things in existence, sons and fathers would be nothing. That's where infinite regress must come to an end with the First Cause. Without the First Cause, nothing could come into existence and remain in existence. Edward Feser uses the striking analogy of IOUs when he speaks of the dog Fido in a linear sequence of dogs. "Just as IOUs have to be backed at some point with real money, so too must a hierarchical series of causes which impart existence to Fido at any moment terminate in something which, since it is Subsistent Existence Itself, needn't have existence imparted to it by anything else."[54]

Third, it doesn't matter whether a series of causes is finite or infinite; in either case, a First Cause is needed. A universe without a beginning and end, for instance, is as much in need of a First Cause as a finite universe. To return to the analogy we used earlier, a chain of causes hanging from an I-beam can be of any length; making the chain infinite doesn't make the I-beam less necessary. A First Cause is necessary for any chain of secondary causes—for otherwise the entire chain would float in the air, until it found a "foundation" to rest on or a "beam" for it to hang from. That's the First Cause, which doesn't have a cause itself—it is uncaused.

If you say that everything must have a cause, then I can only wonder what it is that caused God. If everything has a cause, then God must have a cause too. How can you say that God is uncaused? That sounds quite inconsistent to me.

[54] Feser, *Five Proofs*, 133.

The Argument from Motion

The argument from motion doesn't say at all that *everything* must have a cause. Instead, it says: everything that comes into existence must have a cause. Or even better: every contingent being must have a cause.

Well, God never came into existence. God is not a contingent being but a Necessary Being, so God doesn't need a cause that makes Him come into existence. As a matter of fact, God *cannot* have a cause making Him come into existence, for that would make God a contingent being and rob Him of His power to be the necessary source of all contingent beings. God is the uncaused First Cause of everything else, the unoriginated origin of everything else. The First Cause has inherent causal power, whereas all secondary causes have only derived causal power.

What makes the First Cause so special is that it needs *no* cause. Secondary causes *do* need a cause—a First Cause. Only the First Cause can cause secondary causes to be causes of their own. It is the Cause outside the chain of secondary causes that enables the causes inside the chain. The First Cause is the eternal cause that has always been in existence. Not only is it uncaused, it's not even self-caused—it is pure existence.[55]

With God being a Necessary Being, it makes no sense to ask how God came into being. Just as it makes no sense how eleven came to be a prime number—a necessary mathematical truth—neither does it make sense to ask how a Necessary Being came to be that way. Necessary truths are eternal and unchanging, and so is a Necessary Being. Therefore, the question "Who created God?" is illogical, just as the question "To whom is a bachelor married?" is illogical. It makes no sense to ask "What is God's cause?" because God never began to exist, and only that which begins to exist requires a cause. Those who keep asking this question are basically

[55] *Summa Theologica*, I, Q. 2, art. 3.

asking a nonsensical question: "What is the cause of something that cannot in principle have any cause?"

On the other hand, although God's existence does not have a cause, it does have an explanation, which lies in God's nature—being subsistent Existence Itself. God is self-explanatory, but nothing else is.

The whole argument from motion is based on causes and the existence of causes. But I wonder what happens to the argument if there are events or phenomena that have no cause at all. As a matter of fact, quantum theory seems to suggest the possibility of phenomena without causes.

Let me mention first that there are at least seventeen interpretations of quantum theory, so we shouldn't put all our stock in one of them.[56] One of the most popular interpretations comes from the Danish physicist Niels Bohr, who went as far as to claim that physical systems generally do not have definite properties prior to being measured. The so-called Uncertainty Principle of Quantum Mechanics, for instance, says that an object cannot have both a well-defined position and a well-defined speed: the more accurately one measures the position, the less accurately one can measure the speed, and vice versa. In other words, reality does not have those properties until we measure them.

In Bohr's interpretation, we seem to have lost contact with reality and causality. Bohr is basically saying that reality does not exist when we are not observing it. The problem is, though, that our perceptual experiences give us knowledge of the external physical world only because they are *causally* related to that world. To deny causality in the name of science would therefore be to undermine

---

[56] See, for instance, Graham P. Collins, "The Many Interpretations of Quantum Mechanics," *Scientific American*, November 19, 2007.

the very empirical foundations of science. How could we ever account for our knowledge of the world that physics tells us about if we have no causal contact with it at all? Albert Einstein, for one, always resented and resisted this implication.

Even if we do go with this interpretation, which denies the existence of causes in the quantum realm, we need to realize that quantum events may not have *deterministic* causes, but this doesn't imply that they have no causes at all. Besides, even quantum events still have an explanation based on the laws of quantum mechanics—it is these laws that make quantum phenomena intelligible. But even if certain events do not have a completely determinative *physical* cause, that does not mean, according to the particle physicist Stephen Barr, that these events have no cause whatsoever, for not all causes have to be physical causes.[57]

Where does this leave us? The argument of motion keeps standing tall, perhaps even taller than ever. The universal principle of "nothing that changes can change unless it is changed by something else" remains standing. If anyone denies it, we should tell him to keep searching for what he hasn't found yet. Science can never reject this universal principle, for it needs it for its endeavors. The situation is like the principle of falsification, which we discussed earlier. Counterevidence does allow us to falsify hypotheses or theories, but it can never falsify the principle of falsification itself. Falsification is a principle we need in order to do research; it is not discovered *through* research, but it is presupposed and used *in* research.

We must come to the conclusion that what science cannot prove about God's existence can be done by a metaphysical proof, because it is deductive in nature and uses a universal, self-evident principle as a starting point. This allows us to know with certainty

---

[57] Barr, *Modern Physics*, 264.

that nothing discovered in science could exist if there were no First Cause (or First Mover). We don't need science; we can't even use science to prove the existence of a First Cause. Fortunately, there are better ways to do so, as we found out.

5

The Argument from Design

The argument from design—sometimes also called the teleological argument—has probably been the most influential argument for the existence of God through most of history, but also the most controversial one. It seems to be the hardest case of proving the existence of God as a Designer. It has probably become so controversial because teleology has become more controversial. Teleology embodies the idea that nature has "goals" and "ends," and this suggests that many natural objects appear to have been designed for a purpose: the eye for seeing, the ear for hearing, the heart for pumping, the hand for grasping, and so on.

Aquinas used teleology this way in his argument from design. His reasoning is as follows:

> We see that things which lack knowledge, such as natural bodies, act for an end, and this is evident from their acting always, or nearly always, in the same way, so as to obtain the best result. Hence it is plain that they achieve their end, not fortuitously, but designedly. Now whatever lacks knowledge cannot move towards an end, unless it be directed by some being endowed with knowledge and intelligence; as the arrow is directed by the archer. Therefore, some intelligent

being exists by whom all natural things are directed to their end; and this being we call God.[58]

Aquinas is certainly not a teleological simpleton. He is very well aware, for instance, that plants don't grow toward the light because they know where the light is. As he puts it, they "lack knowledge." What, then, does explain the fact that plants grow in the direction of light, if it's not knowledge? Aquinas's answer to this question is that they must be "directed by some being endowed with knowledge and intelligence"—that is, designed to be such as to grow toward the light. That's a teleological explanation.

The argument of design as a teleological argument was "rejuvenated" by William Paley at the end of the eighteenth century with his famous "watchmaker" analogy. Paley argued that the beautifully designed works of nature must have had a Designer, just as a watch must have one.[59] Paley's teleological argument could be laid out as follows:

Human artifacts are products of intelligent design.

The works of nature resemble human artifacts.

Therefore, the works of nature are a product of intelligent design.

In Paley's words:

In crossing a heath, suppose I pitched my foot against a stone, and were asked how the stone came to be there; I might possibly answer, that, for anything I knew to the contrary, it had

---

[58] *Summa Theologica*, I, Q. 3, art. 2.
[59] William Paley, *Natural Theology, or Evidences of the Existence and Attributes of the Deity collected from the Appearances of Nature* (Oxford, UK: Oxford University Press, 1802, 2006).

lain there forever: nor would it perhaps be very easy to show the absurdity of this answer. But suppose I had found a watch upon the ground, and it should be inquired how the watch happened to be in that place.... There must have existed, at some time, and at some place or other, an artificer or artificers, who formed [the watch] for the purpose which we find it actually to answer; who comprehended its construction, and designed its use.... [Every] manifestation of design, which existed in the watch, exists in the works of nature; with the difference, on the side of nature, of being greater or more, and that in a degree which exceeds all computation.[60]

The approaches of Aquinas and of Paley are both teleological. Aquinas believed that everything in the universe has a goal and that this goal is given to it by God, just as the arrow flying through the sky is given its goal by the archer who fires it.[61] Paley believed something similar: just as watches, which exhibit complexity and purpose in order to enable us to know the time, have watchmakers, so must the works of nature, which have complexity and the goal of sustaining life, have a Maker too.

It surprises me that the argument from design, unlike the other proofs of God's existence, doesn't start with a universal principle, independent of sensorial experience. Instead it uses the analogy of archers, watches, and the like. Analogies are always questionable. The idea behind the analogy of a watchmaker is that the world of living beings and the world of technology share the complexity of design, and therefore most likely also need a Designer of this design. But what is this analogy worth, and, in general, what is any analogy worth as a proof of God's existence?

---

[60] Paley, *Natural Theology*, chap. 1.
[61] *Summa Theologica*, I, Q. 2, art. 3.

The use of analogies such as the watchmaker in the teleological argument does make things more comprehensible to us: it uses something in our experience to try to explain something beyond it. The role of analogies in a proof of God's existence is very limited, though.

The argument runs like this: things such as watches display design; these things were designed by watchmakers; therefore, everything else in this world was designed too. But it does not follow conclusively that everything else must have been designed and had a Designer as well. The things being compared may not be similar enough for this argument to apply. To put it simply, not all things are like watches; we cannot reason safely from a maker of watches to a Maker of all there is. All arguments by analogy rely for their strength on the similarity of the things being compared. They may not be comparable in having a designer.

It can be questioned, though, whether someone like William Paley really uses the analogy of a comparison between the machineries we produce and the "machineries" we find in nature. Paley considers all machineries in this world—human or natural—of the same kind. As he puts it, "[Every] manifestation of design, which existed in the watch, exists in the works of nature." Paley considers living systems no different from artificial systems; therefore, the same effects—working machineries—must have the same causes—a design and a designer. For Paley, the eye is a machine, as a telescope is—they are both instruments. This explains how Paley can speak of *"precisely the same proof* that the eye was made for vision as there is that the telescope was made for assisting it."[62] Interestingly enough, he never likened the universe to a watch, nor God to a watchmaker.

Paley's argument is not based on our knowing that watches are designed because we have seen them being designed and that

---

[62] Paley, *Natural Theology*, chap. 3 (emphasis added).

objects in the universe are watch-like and thus are likely to be designed too. What he argues instead is that even if we had no knowledge that watches were made by human beings, we could still infer that a watch was designed, because such a finely tuned, goal-serving structure could not have originated from any non-goal-directed process.

I think the real problem of the teleological argument is not so much the use of analogy as it is its teleology, which speaks of designs that have goals and purposes. The watch may have a perfect design made for the purpose of showing us the time, but what makes us think that all phenomena in nature — not only the technological pieces — have a goal or purpose? Didn't Charles Darwin ban the "design" concept once and forever from biology? George Bernard Shaw once said that Darwin had thrown William Paley's "watch" into the ocean.[63]

Did he really? I wonder whether he did. I agree with you that most biologists are very reluctant to use teleological statements. Many scientists, especially biologists, have become increasingly "teleophobic." A modern representative of this view is the biologist Richard Dawkins. He has been referred to in the media as "Darwin's Rottweiler," a reference to English biologist T. H. Huxley, who was known as "Darwin's Bulldog." Although Dawkins sees in evolution "the appearance of having been designed for a purpose," he considers that a deceptive appearance. In his own words: "So powerful is the illusion of design, it took humanity until the mid-nineteenth century to realize that it is an illusion."[64] Darwin's Rottweiler thanked Darwin for this awareness.

---

[63] Marjorie Green, *The Knower and the Known* (Berkeley, CA: University of California Press, 1974), 195.

[64] Richard Dawkins, "Big Ideas: Evolution." *New Scientist* 14 (September 2005).

Unfortunately for our discussion, "goals" and "purposes" in nature are often associated with "intentions" that human beings can have. Eye patterns on butterfly wings have the effect of keeping predators away, but that's not an intention butterflies have "in mind." So, in a sense, you're right: the sun, for example, does not rise every morning because it wants to. Water does not seek its own level because it intends to. Eliminating intentions in biological explanations, however, does not entitle us to eliminate goals and purposes as well. Many things in the living world do have a certain goal or outcome, but without being intended. They happen *for* a specific purpose or goal or end.

Here are some examples. The heart is for blood circulation. Hemoglobin is for the purpose of transporting oxygen and food. The green color of caterpillars is for camouflage. The eyes on butterfly wings are for deception. The eye is for seeing, the ear for hearing, the hand for grasping, and so forth. Not to make connections like these would be silly. As to how these features came to be is an entirely different question, but that does not entitle us to reject teleology itself. Biologists might like to do so, but they cannot live without teleology, and they usually express their findings in teleological terms. Most of their books and articles would have to be rewritten if teleology were anathema. In short, there is no good reason for "teleophobia" in biology.

Let's revisit the question: Did Charles Darwin discard the word *design* from biology? I think that is a misperception of Shaw, Dawkins, and many others. After Darwin, the heart still existed for circulation; the cause of its existence may have been different, but its teleology was not.[65] As a matter of fact, Darwin put

---

[65] This point was made very convincingly by Michael Ruse, "The Last Word on Teleology," essay 4, in his *Is Science Sexist?* (Boston: Reidel Publishing, 1981).

purposes—more neutrally referred to as *functions*—back into the picture. Apparently, Darwin did not discard design or what comes with it, teleology. The concept of *design* is an artifact analogy that is as basic to Darwin's theory of evolution as it is to Paley's natural theology. The heart has a design as much as a pump has a design. They both have a design fit for pumping. How such a feature came along is another story.

As a matter of fact, each time evolutionary biologists speak about natural selection as selecting what is "fit" or "successful," they are talking teleology based on the notion of design. All causes have effects, but certain effects are more successful for survival than others. Certain biological features of organisms are "successful" and "effective" in reaching their "goal" because they have a design that enables such a successful outcome. If these features were not design-like, they simply would not work. Leon Kass, a University of Chicago professor and physician, could not have worded it better: organisms "are not teleological because they have survived; on the contrary, they have survived (in part) because they are teleological."[66]

In other words, natural selection does not create the fit—it only selects what fits. It does not explain a fit but uses a fit in order to select. So the fittest are not defined by their survival—that would make for a tautology—but by their design. Different designs have different effects that makes them "successful" to different degrees. Consequently, biological fitness is not an outcome of natural selection but a condition for natural selection. We are inevitably talking teleology here.

Darwin may indeed have thought he could reduce teleology to the causality mechanism of natural selection, but this mechanism

---

[66] Leon R. Kass, M.D., "Teleology, Darwinism and the Place of Man: Beyond Chance and Necessity?" in *Toward a More Natural Science* (New York: Free Press, 1988), chap. 10.

can work only on condition that there is teleology in nature. Even a process such as natural selection is based on a design-like world. The results must be teleological and design-like, because if they were not, they simply would not work in solving the problems they face during the process of natural selection. If the eye lens, for example, did not function like a physical lens, one would not see very well. Teleology in the context of natural selection merely means that causes have effects, and that some effects are better than others for survival and reproduction.

Natural selection may explain that a fine working design has a better chance of being reproduced, but ultimately it cannot explain why such a design is working so well. And that's where teleology is needed—even in Darwinism. In that sense, Darwin did not change teleology from an "a priori drive" into an "a posteriori result." Teleology is not a biological outcome "a posteriori," but rather a metaphysical given "a priori." Natural selection does not create teleology, but its working is based on teleology. There is "something" in successful biological designs that carries them through the filter of natural selection.

This makes me think that the real problem of the teleological argument is probably located in the notion of design itself.

Indeed, the word *design* is very ambiguous and not well defined. It is important to notice that the word *design* in English can refer not only to purposes but also to patterns (or to both). If we connect design in nature solely with purposes, we may limit ourselves to the animate world. If, on the other hand, we use it to refer to patterns, we open up the inanimate world as well.

Taken in the sense of a pattern, *all* of nature is permeated with design. There is as much design in the inanimate world as there is in the animate world. What kind of pattern is this? It is the pattern of *regularity*—the regularity of "like causes having like effects." In the words of Aquinas, "Things ... act for an end, and this is evident from their acting always, or nearly always, in the same way."[67]

If there were no regularity in the universe, it would make no sense to search for laws in physics, chemistry, or the life sciences. It is only because we "know" that like causes produce like effects that we are able to explain and predict in science. The laws of nature that science brings to light are the many patterns of regularity we detect in this universe. In other words, regularity is proto-scientific and must come first, before science can even get started. Science in an irregular, chaotic world would not make any sense. G. K. Chesterton once "seriously joked," as only Chesterton can, about a conspiracy of order in our world of regularity: "One elephant having a trunk was odd, but all elephants having trunks looked like a plot."[68]

Regularity is part of what could very well be called the "cosmic order." Science can help us to reveal this cosmic order and investigate its details. Especially in physics, empirical laws have been found to follow from deeper and deeper laws and principles. Chemical bonds, for instance, follow from laws of atomic physics, and these, in turn, flow from the laws of quantum electrodynamics. This is how deeper and deeper levels of laws have been uncovered. As Stephen Barr puts it, "The deeper one goes the more orderly nature looks, the more subtle and intricate its designs."[69]

Does this legitimize the use of the term *design*? Unfortunately, the concept of design is often seen as a threat to science because it seems to point to a Designer. As a result, the argument from design is considered suspect in science, and scientists feel forced to avoid

---

[67] *Summa Theologica*, I, Q. 3, art. 2.
[68] *The Essential Gilbert K. Chesterton* (New York: Simon and Schuster, 2013), 43.
[69] Barr, *Modern Physics*, 81.

the term *design*. The term itself, however, is rather neutral; it may point to a Designer, but it does not necessarily prove the existence of a Designer. Science doesn't have that irrefutable power. Only proofs of God's existence can achieve that certainty. So that would be a task for the argument from design.

As I said earlier, unlike the other proofs of God's existence, the argument from design doesn't start with a universal principle. Instead, it uses analogies, which are always questionable by their nature. As a result, the argument from design is inductive rather than deductive, and therefore does not lead to a conclusive proof of God's existence.

That's true—the argument from design lacks the rigor of the other proofs of God's existence that we have discussed so far. It is very well possible, however, to come up with a universal statement that starts the proof. It would be something such as "Like causes have like effects." Causes and effects are related to each other in a lawlike, regular manner.

Some might say this statement also requires some sensorial experience, but that is not quite true. If experience seems to tell us that "like causes" do *not* have "like effects," then we assume they must not be "like causes," or they must not be "like effects." If the cause is the same, the effects must be the same; if effects are different, then the causes must be different. So the statement that like causes have like effects precedes experience and observation and isn't based on experience and observation. This universal principle is so self-evident that even scientists accept it as a given. They assume there must be a law-governed connection between causes and effects. Starting with a statement like this, the argument from design could become deductive instead of inductive.

Seen this way, the argument from design could have the following structure:

Every contingent being has a design of "like causes having like effects."

Such designs cannot explain their own existence, so there must be a First Designer.

The First Designer must be a Necessary Being and therefore doesn't need a designer Himself.

Based on the premise of a universal principle, the conclusion follows with necessity. With a structure like this, we have another proof of God's existence.

Some might object that the universal principle used here does not apply to the *goals* that the argument from design is about. But that's not quite true either. It could very well be argued that the effects caused by certain causes are the "goals" of those causes. For Aquinas, there is no way to make sense of the fact that cause X always generates effect Y—rather than Z, or no effect at all—if we don't suppose that X always points to or is always directed at Y as toward an end or goal. All causes are goal-directed or effect-oriented, so to speak. Their causal power goes hand in hand with goal-directedness; deny the latter, and you implicitly deny the former. Their causal power is derived, however, whereas God's causal power is inherent and doesn't derive from anything else.

Even if you are right that Charles Darwin did not throw Paley's "watch" into the ocean, he certainly did throw Paley's Watchmaker out the window—including the argument from design.

Whether Darwin did so or not, he at least thought he did. He strongly believed that his causal mechanism of natural selection could effectively replace the need for a Cosmic Designer.

What Darwin and most of his followers don't seem to realize, however, is that the Designer does not *replace* secondary causes but causes them to be secondary causes of their own—which would include the mechanism of natural selection as well. So even if there is evolution, this doesn't leave the Designer with nothing to "do"; it actually makes a Designer more needed than ever. There is no competition, let alone a conflict, between what the Designer does and what natural selection does. They are not replacements for each other, but work in tandem.

As early as 1868, John Henry Cardinal Newman was very aware of the harmony between these two accounts when he wrote in one of his letters, "Mr. Darwin's theory need not then to be atheistical, be it true or not; it may simply be suggesting a larger idea of Divine Prescience and Skill."[70] He also wrote in that same letter, "I do not [see] that 'the accidental evolution of organic beings' is inconsistent with divine design—it is accidental to us, not to God." The divine design explains why the world is the way it is. Stephen Barr put it very pithily: "We now have the problem of explaining not merely a butterfly's wing, but a universe that can produce a butterfly's wing."[71]

The fact that there is a cosmic order does not mean our explanations can go endlessly deeper and deeper. That would amount to digging oneself into a deeper and deeper hole. Going deeper and deeper leads to a regress of laws of the *hierarchical* type that keeps floating in the air if the sequence of laws and explanations has no "basis" to rest on or no "hook" to hang from. This means that the regress must terminate somewhere—not in another "ultimate" law, but rather, in something that is self-explanatory. Let's not forget that the pattern of regularity on its own cannot be the final

---

[70] John Henry Newman, "Letter to J. Walker of Scarborough, May 22, 1868," in *The Letters and Diaries of John Henry Newman* (Oxford: Clarendon Press, 1973).

[71] Barr, *Modern Physics*, 112.

explanation. Because the cosmic design is contingent—it could have been different—and therefore is not self-explanatory—it needs an ultimate explanation beyond itself.

This calls for a Cosmic Designer. God not only brought the universe into existence; He also keeps it in existence. The ultimate cause of "all like causes having like effects" is a Divine Lawgiver, a Divine Designer. Aquinas concludes, "Therefore, some intelligent being exists by whom all natural things are directed to their end; and this being we call [a] God."[72] We have reached here the end of the argument from design. It no longer makes sense to ask who designed the Designer. The Designer does not need to be designed, for He is the origin and source of all design. In short, God is an undesigned Designer.

The Argument from Universals

This proof of God's existence is rather different from the previous ones, but it is one of my favorites. It starts from concepts and then reasons toward the existence of God.

You may not realize it, but concepts are central to human life. Whatever we do—talking to other people, thinking about certain issues, doing research, having discussions with other people, or studying proofs of God's existence—we are using *concepts*. These concepts feature in statements or propositions. When I make the statement "Snow is white," for example, I use at least two concepts: the concept of *snow* and the concept of *whiteness*. For human beings, it's hard to imagine life without concepts. That may sound obvious until we begin to ask ourselves what those concepts really are.

The concept of *concepts* is rather abstract and ethereal. Yet concepts are essential as the building blocks of statements or propositions.

[72] *Summa Theologica*, I, Q. 3, art. 2.

A Catholic Scientist Proves God Exists

There are simple statements, such as "Snow is white"; more intricate ones, such as "Litmus paper turns red in acids"; rather sophisticated ones, such as "Genes are made of nucleotides"; and more complicated ones, such as "Radioactive isotopes decay with a constant half-life rate." No matter what the statement says, we don't understand what it says if we don't understand the concepts it is based on.

Concepts are *universals*: they abstract from particular observations that which is universal. Concepts have the universality that observations miss. Whereas material things are always particular, concepts are always universal. From this we may conclude that concepts, and the propositions that contain them, are *abstract* objects. Indeed, the very concept of *concepts* couldn't be more abstract and ethereal. Let's delve a bit deeper into the status of concepts.

I don't think we need to. How could concepts be anything else than mere thoughts? I would say that the idea that concepts are just thoughts is hard to deny.

I'm not so sure. Concepts cannot be mere thoughts. The reason is basically very simple: If concepts were just thoughts, then we could not communicate with each other, because I cannot read your thoughts, and you cannot read mine; I wouldn't know what you were thinking when I used certain concepts for something I was thinking of. If we were to speak about something such as a gene or a mutation or any topic, I would never be able to know you were thinking the same thing I was thinking. When you and I think about randomness, for instance, it is not that both of us are entertaining our own private thoughts of randomness with nothing in common between us. If that were the case, each one of us could have a different thought or idea of randomness, and real communication between us would be impossible.

When the two of us have the same thought, however, there must be something we have in common that transcends our private thoughts. That's where concepts come in. Even when I write my thoughts down, I need concepts to do so. In other words, we may think about a concept, but the concept itself is not a thought — it's only the object of a thought. Concepts do not come from thoughts, but they make thoughts possible. According to St. Augustine, there must be something "that all reasoning beings, each one using his own reason or mind, see in common."[73]

If concepts are not mere thoughts, then they must be purely linguistic entities. What could be wrong with that solution?

Concepts cannot be merely linguistic entities. It is true that words are linguistic items, but concepts are not — they are extralinguistic entities. Therefore, we shouldn't confuse words with concepts. A word is at best a label for a concept. The word *snow* would be a label for the concept of snow. A concept such as snow can be conveyed by different words in different languages — for example, the word *snow* in English and the word *Schnee* in German. Beneath the different words of different languages, we find common concepts — and this is what makes translation from one language to another possible.

The concept itself would exist, however, even if those languages did not exist or no one had ever used any words to refer to the concept of snow, for instance. We can even talk about snow existing at a time when there were no human beings around yet. Claiming that the term *white* "in "Snow is white" is just a term and that there is no such thing as whiteness should be countered by asking why we apply the term *white* to just the things we do. It's hard, if not impossible, to come up with a better answer than "because they all have whiteness in common" — and this brings us back to affirming the existence of universal concepts after all. We

[73] Augustine, *On the Free Choice of the Will*, 2.8.20.79.

may talk about concepts with words, but the concept itself is not a word. Put more generally, concepts do not come from language; instead, they make the use of language possible.

If concepts are not mere thoughts or mere linguistic entities, they must come from definitions that we make up to explain what they stand for. I don't see any other solution.

Sorry, but the problem of the solution you suggest is that definitions inevitably require other concepts. Take the following references in a dictionary: "A gene can take alternative forms called alleles," and "An allele is one of the alternative forms of a gene." These two statements are obviously circular references that explain a concept with the very same term it is supposed to explain. Even if the definition is not that trivial but provides a more extensive description, the circularity remains, for every description of a concept requires the use of other concepts, which, in turn, require more and more descriptions using concepts. That's an unending task.

Dictionaries, for instance, must ultimately always use circular reference, since all words in a dictionary are defined in terms of other words. A dictionary can never step outside its confines to refer to something outside the dictionary. No matter how hard we try, we will never be able to get the concept we try to describe and define "off the ground"—every try falls back on other concepts. Even if we decide to stop this endless regression by declaring a few concepts as the pillars that carry the rest of the conceptual framework, we still need to explain where those fundamental concepts come from.

If concepts do not come from definitions, they must come directly from observations. I can explain the concept of snow by pointing at the snow we see. I don't see how you could ever deny the obvious.

Well, you could be wrong again. Take the concept of hydrogen. It does not arise from particular observations, because no observation can tell us whether a certain gas is hydrogen until we know what hydrogen is and how to test for it. Besides, hydrogen is not visible. Or when we see blood stains at a crime scene, the mere observation of those stains doesn't make it blood. Forensic scientists will tell you they need something such as luminal to determine if it is blood, as luminal reacts with the iron in hemoglobin.

But what is even more crucial is this: observation is always about particular things, whereas concepts are universal. As we said earlier, concepts abstract from particular observations that which is universal. Concepts have the universality that observations miss. Material things are always particular, whereas concepts are always universal. We may have seen many circular objects, but we have never seen the perfect circle that the concept of circle is about. Even the concept of uniqueness is universal. It does not apply to one single, unique case but puts a thing into the category of things that are unique in one or more respects. It's a universal concept about many concrete cases that are similar in a certain aspect—that is, in being unique.

If observation can't do the job, what else could? I see only the possibility that concepts are established by pointing at what the concept refers to.

That, too, may be a hard case to make. First of all, for certain concepts, there may be nothing to point at. To explain the concept of *tomorrow*, for instance, there is nothing to point at (other than on a calendar, but that requires the concept of calendar as well). Neither does the mathematical concept of pi (π) refer to any object in the world that we can point to. There are no pi objects in the world.

Second, pointing at a cell under the microscope, for instance, does not automatically generate the concept of *cell*. The history of cell biology shows us how tedious its discovery was. Even showing or pointing at many cells does not generate that concept. Claiming that objects are alike in a certain respect—for example, being a cell—invokes another abstract and universal concept, *similarity*, which leads us into a vicious regress problem. To decide which objects are to be included in the set and which are not, we need a criterion that says that only cells are to be included in the set. This account presupposes the very concept of *cell*, the acquisition of which it is meant to explain.

It is only when we know the concept of *cell* that we can point at a cell and identify it as a cell. Anything can be pointed at once it has been identified; but not everything that has been pointed at can be identified with a concept. Just ask biology teachers, who must teach their students to "see" cells under the microscope—it's not as easy as it sounds. As St. Augustine put it in *On the Free Choice of the Will*, "What is recognized is present in common to all who recognize it." Put differently, there must be some cognition before there is recognition.

Perhaps there may be nothing material to point at to establish a concept, but the concept itself may be a material entity—something located outside the mind that we can point at.

I am afraid your suggestion cannot be true either. As we said earlier, material things are always particular, whereas concepts are always universal. The concept of *circularity*, so we found out, is not about one particular object with the shape of a circle, but instead it refers to *all* objects with such a shape. The concept of *circle* can be used for any specific circular object, regardless of its size and its imperfections. Concepts have a universality that material objects can never possess. Obviously, there is no material entity outside the mind that the concept of *circle* refers to.

Besides, even if a concept does refer to a material entity, the concept itself is not a material entity existing outside the mind in the world around us. Concepts don't have material properties—they have no weight, no shape, no color. Therefore, the concept of *circularity* is not circular itself; the concept of *electron* is not an electron itself; the concept of *neuron* is not a neuron itself. All particular entities in the world around us, outside our minds, are material, whereas concepts are nonmaterial.

If concepts cannot be material entities located outside the mind, they might be material entities located *inside* the mind.

But this idea cannot be true either. To show why not, I may have to explain a bit more, as it is a rather popular misconception among scientists.

It seems to be very tempting for scientists to get rid of the highly abstract and ethereal notion of concepts by reducing them to something that is the product of neural activity in the brain. The famous example to defend this position is the following: When a frog sees a fly zooming by, the frog's brain displays a certain pattern of neural firing. In a similar way, when we see a tree, there must be a distinctive pattern of neural firing in our brain that is correlated with and caused by seeing a tree. This has led some to believe that thinking of a certain concept—say, a *fly* or a *tree*—is also and only a certain pattern of neural firing. The problem here, however, is that seeing a fly or a tree is a matter of perception, whereas using a concept is a matter of thinking. Thoughts have meaning and content; perceptions do not.

If my thoughts about mathematical circles, for example, were just a physical representation in the form of a neuronal firing pattern somewhere in the brain, those thoughts would at best be another particular material thing. But this necessarily means that those physical, material representations could not be universal. Reducing concepts to material entities—say, neuronal firing patterns in the brain—would make them something particular. Particular material things cannot qualify as universal. If the thought about circles were indeed a particular neuronal firing pattern, that pattern itself would have to be a circle as well; this is obviously nonsense.

Besides, to reduce concepts to a "product of neurons" obscures the fact that *neuron*, too, is an abstract concept. That would make for a pernicious vicious circle: the very idea that concepts are nothing but neurons firing is itself nothing but neurons firing.[74] Those who claim that concepts are merely products of neurons should realize that talking about neurons requires the nonmaterial concept of *neuron* to begin with. In other words, concepts do not come from neuronal activities; instead, we can understand neuronal activities and talk about them only with the help of certain concepts—in this case, the concepts of *nerve*, *neuron*, *synapse*, *neurotransmitter*, and the like.

If it is true, as you say, that concepts and propositions don't come from our thoughts, or from our observations, or from our languages, or from our surroundings, or from our brains with their neurons, then we should ask ourselves where else they could come from. They must exist somewhere for them to be available to each one of us and all of us together. But where is that somewhere?

If abstract objects such as concepts and propositions do not depend for their existence on the material world or on the human mind—on something contingent, that is—then there is only one rational option left: they must exist in a third realm that is not contingent, neither material nor mental.

This idea is usually associated with the philosopher Plato, who mentioned this third realm long ago. Plato's position faces several, rather technical problems, however, which we will not discuss here. But there is a much more acceptable version of this third realm—arguably the only valid one—which goes basically back to

[74] Barr, *Modern Physics*, 216.

St. Augustine[75] and was later elaborated on by the philosopher Gottfried Leibniz.[76] This version holds that abstract objects do indeed exist, but they can do so only in an *infinite, eternal, Divine Intellect.*

That step goes too fast for me. Why can't abstract objects exist in human intellects, instead of a *Divine* Intellect?

The reason is that human intellects are contingent. They do not have to exist, but they come into being and pass away. If abstract objects existed only in human intellects, they would have to come into existence and could go out of existence too. In addition, we could not have them in common with other human intellects. And more important, before humanity emerged, including human intellects, there would and could not have been any abstract concepts if they had to come with human intellects. Snow, for example, was already white before human intellects emerged.

So the only sort of intellect on which abstract, universal, and timeless concepts and propositions could ultimately depend for their existence would be an intellect that could not possibly have *not* existed, but exists in an absolutely necessary way—which is a Divine Intellect, the Mind of God, the First Cause. It is this Divine Intellect that grasps and holds all of the logical relationships between all propositions with all their universal concepts—a being who eternally understands all actual truths, plus all possible truths, as well as all necessary truths. For example, a proposition such as "It is true that snow is white" exists in the Divine Intellect. A proposition like this is either true or false because God causes the world to be such that this proposition is either true or false. Therefore, propositions are true, because they exist as true thoughts in the Divine Intellect.

---

[75] Especially in book 2 of *On Free Choice of the Will.*
[76] Especially in sections 43–46 of his *Monadology.*

A Catholic Scientist Proves God Exists

Similar to what the First Cause argument does, the eternal-truths argument proves that the truths of concepts and propositions can exist only within a Divine Intellect and could not exist if God did not exist. Their existence cannot be grounded in contingent entities, so they must find their ultimate ground in a necessarily existing intellect. No wonder, in his Fifth Way, Aquinas establishes that God has intellect.[77]

Although the "eternal truths argument" the way you describe it is in essence a philosophical proof of God's existence, I don't see how it could square with what religion says about God.

I think the two are perfectly consistent with each other. The Judeo-Christian faith tells us that we were made in God's likeness and image.[78] Since we were made in God's image and likeness, our finite human intellects are a reflection of God's Divine Intellect and have some kind of access to that Divine Intellect. Our human intellect is somehow able to capture concepts and propositions that reside in God's Divine Intellect. Only God's Divine Intellect can make it possible for you and me to share the same concepts and propositions when we say and think that snow is white. And the same holds, of course, for all the other concepts and propositions that we entertain, including the ones from science and religion.

To put it in a more charged way: without faith in God's Divine Intellect, we have nothing to claim as truth. We are entitled to say that the statement "snow is white" is true only if snow is indeed white in the mind of God. One of the ways to find out whether it is indeed in the Intellect of God is by "reading God's Mind" in nature and through reason. We do so, for instance, when we "interrogate"

---

[77] *Summa Theologica*, I, Q. 14, art. 1.
[78] See Gen. 1:27.

the universe by investigation, exploration, and experiment, but also by using logic, reason, and philosophy.

You explained that concepts must reside in God's Intellect, but you talk also about propositions residing there. The term *proposition* seems technical. It is not clear to me why you don't call those propositions simply *statements*, sentences that state what the facts are.

That would not be a good idea, because the distinction between sentences and propositions is quite essential. Here is why. If they were the same, we would run into the logical problems of paradoxes—which indicates there is something wrong with them. Indeed, the sentence "Snow is white" is true if it corresponds to the fact that snow is indeed white. This, however, could lead to a paradox. Think of this simple paradox: a card, on one side of which is written "The sentence written on the other side is true," while on the other side is written "The sentence written on the other side is false." One could easily arrive at the paradoxical conclusion that either sentence on the card is both true and false at the same time.

The logician Alfred Tarski showed us how to avoid a paradox like this.[79] He told us to decide whether the sentence that was written on either side of the card is within the language system in which the talking is being done or within the language system being talked about. If both sentences are taken to be in the language system that is being talked about, then they cannot also be taken as referring to each other. In situations like these, we have two expressions: one is a sentence, the other one is a proposition. Given this distinction, we can now say that the sentence "Snow is white" is true if it corresponds to the proposition that snow is white.

[79] Alfred Tarski, "The Concept of Truth in Formalized Languages," in *Logic, Semantics, Metamathematics* (Indianapolis, IN: Hackett Publishing, 1983), 152–278.

A Catholic Scientist Proves God Exists

Another reason why sentences must be distinguished from propositions is the following. Propositions are truth bearers—they are either true or false. They are distinct from the different sentences we might use to express them. "Snow is white" and "*Schnee ist Weiss*" are different sentences in different languages, but they both express the same proposition. That's why two people can have the same thoughts, even when they use different languages with different sentences. You can make a sentence visible by writing it on paper, but the proposition it expresses is invisible. Like concepts, propositions are abstract objects—objects of thought, yet independent of thought.

Sentences are "real"—they can be spoken, heard, read, or written. But I wonder how "real" those invisible propositions are.

There are many reasons why propositions and other objects of thought are very real. The most important argument is that propositions do not depend on human brains, or even minds, for their existence. The facts contained in propositions exist on their own, even when no human being is aware of those facts. The proposition "Snow is white" is true, regardless of our considering it or expressing it in a sentence; it was already true even before humanity appeared on Earth. When two scientists are thinking about the law of gravity, they are both thinking about one and the same truth—which is the truth of the proposition "Gravitation varies with the square of the distance." That proposition was already true before Isaac Newton came up with gravitation.

Where do propositions like these come from? As we found out earlier, they can exist only in the Divine Intellect. Sentences come from the human intellect, but propositions come from the Divine Intellect. The existence of a Divine Intellect would also explain why there seems to be a rather perfect "match" between the rationality of our minds and the "rationality" found in the world around us—which we called earlier the Cosmic Order. There is some kind of match between thoughts and reality. Somehow the human intellect seems to be able to capture reality the way it *is*.

Not only does the physical order we observe in this world appear to be amazingly "consistent," but so does the world of thoughts in our minds. It is a consistency that should perplex us. If we deny the match between these two—if no connection exists between what is out there and what is in our minds—then all we know or can know is our thoughts. If our minds have no connection with things, we can formulate whatever world we would like to have. The disastrous outcome would be that our minds would have no reality check any longer. We would have lost contact with our world. That outcome would be detrimental to us if it were true.

One way of summarizing the argument from universals in an abbreviated form goes as follows:

All concepts are abstract, universal entities.

These concepts cannot be based on contingent intellects.

Therefore, they must reside in a necessary Divine Intellect.

Conclusion

All the proofs we discussed in this chapter turned out to be of the deductive type. They prove to us that God is Existence Itself, a Necessary Being, the First Cause, the Cosmic Designer, and an Eternal Intellect. Only this God is an answer to the question of why there is something rather than nothing—or better: why there are any contingent things at all. In answer to that question, only the God of the proofs could possibly terminate the regress of causation and explanation. Science cannot really help us here, because it limits itself to secondary causes, whereas the question of God's existence is about the First Cause.

A Catholic Scientist Proves God Exists

This gives the proofs of God's existence strong deductively conclusive support, whereas science can provide only inductive evidence. It also sheds a powerful light on many questions we have. How, for instance, can we explain our existence? In a linear sequence, we may go back to previous ancestors. In a hierarchical sequence, we may go into biological causes, which we then can trace back to chemical causes, which, in turn, can be derived from physical causes. But all these causes combined don't really explain anything. For the entire chain of these causes is just floating in the air, and it keeps doing so until we find a "foundation" for it to rest on, or a "beam" for it to hang from. That's why God's existence is so vital for our existence.

It's important to notice that these proofs are based on common sense and principles that are accessible to everyone, without reference to any specific religion. Anyone who seriously studies these arguments must come to the conclusion that God's existence has been proven in the strict sense. The conclusion always follows necessarily from the premises—if the premises are true, the conclusion must be true too. It is impossible for the premises to be overthrown by further empirical inquiry, for these premises are a precondition for any empirical inquiry to be even possible.

If the proofs you explained above are really as certain and convincing as you claim they are, how come there are still people who do not believe in God, let alone in the proofs of God's existence? Shouldn't everyone have been "converted" from disbelief in God's existence to belief in God's existence?

There may be several reasons why that hasn't happened. First, the fact that someone somewhere has raised an objection against an argument for God's existence does not prove that the argument fails. This is a false conclusion, which would make an argument valid only if no one objects to it. That argument is not only absurd; it is never accepted as refuting other claims to which at least some people object (such as that free will exists or that science is the only source of knowledge or that other persons exist).

Second, many people, especially some scientists, have attacked proofs of God's existence that they don't properly understand. They misinterpreted them before they attacked them. They have never taken the time and energy to study the proofs carefully. Thus, they are actually attacking a straw man that they borrowed from others or from someone who started the misconception. One of the most common misunderstandings can be found in the use of the incorrect universal statement "Everything has a cause." We've pointed out this error repeatedly. None of the proofs is about "everything," but only about "everything that comes into existence." But I think we have said enough about that point.

Third, even if a proof is conclusive, that doesn't mean everyone will accept it as conclusive. What we often see in the lives of nonbelievers is a certain disconnect between what they know and what they choose to know. St. Thomas Aquinas was aware of this disconnect: "Whereas unbelief is in the intellect, the cause of unbelief is in the will."[80] No matter how strong the evidence is in favor of God's existence, some atheists *prefer* not to accept God's existence as a fact because they don't like the way the world looks to them with God in the picture. They act like people who deny the Holocaust and ignore the overwhelming evidence for it because they are not willing to believe it. It is the will, says Blaise Pascal, that "dissuades the mind from considering those aspects it doesn't like to see."[81]

---

[80] *Summa Theologica*, I, Q. 2, art. 2, 10.
[81] Blaise Pascal, *Pensées*, 2, 99.

A Catholic Scientist Proves God Exists

This takes us back to what St. Thomas Aquinas says: "We must believe that [a] God exists, which is clear by reason."[82] What Aquinas means by this statement is that reason leads us to assert God's existence. This does not mean that reason can fully fathom God, for elsewhere he says, "We cannot understand what God is, but what he is not."[83] In other words, our (philosophical) knowledge is limited to knowing *that* God is, more so than *what* or *who* God is. But we do know, based on this, that God must be necessary, unchanging, and infinite—all-knowing, all-powerful, and all-present.

The question remains, though, as to whether the God you are talking about is also unique. Could there be two first causes, two necessary beings, and so forth—with one being a material entity like matter and the other a nonmaterial entity? Or could there be many gods in the way that the Greeks and Romans once thought there were many, with each one in charge of a specific domain?

The answer to this question is a definite no. If there were more than one god—let's say two—then something in them would have to be the reason for their difference.[84] The one possessing the distinguishing addition would not be a god at all. It would be impossible for each of these two gods to enjoy perfectly independent self-existence. One of them would have to give being to the other, or else they would both derive their being from something else, so neither of them could be a god.[85] This makes the existence of more than one First Cause impossible. If there are more, none of them would qualify to be called God.

---

[82] Aquinas, *Compendium Theologiae,* 3.
[83] Aquinas, *Summa contra Gentiles,* I, 30.
[84] *Summa Theologica,* I, Q. 11, art. 3.
[85] For more details, see Augros, *Who Designed the Designer?,* 52–54.

Another way of proving this impossibility goes as follows. When we say that God is infinite, we mean that He is unlimited in every kind of perfection, or that every conceivable perfection belongs to God in the highest conceivable way. This would be impossible if two or more infinite beings were to exist, for each one would have some perfection not possessed by the others. But with an infinite, perfect being, there could not possibly be a way to distinguish one from the other.[86] So neither one of them could be God.

And here is a third way: since God is that being who is just being and existence itself, and in whom essence and existence are identical, there is no sense to be made of the idea that God shares an essence with anything else or has one act of existence alongside others.[87] In other words, there is no logical proof for polytheism, but there is one for monotheism.

---

[86] *Summa Theologica*, I, Q. 11, art. 3.
[87] Ibid.

6

Why God Is a God

In chapter 2, we mentioned that St. Thomas Aquinas is very careful when he says at the end of each proof of God's existence, "And this all think of as [a] God." As said earlier, in his Five Ways of proving God's existence, Aquinas does not use *god* as a proper name but as a common noun. Each one of his proofs concludes only that there is "*a* god." At the same time, Aquinas tells us these proofs are also proofs of the existence of the very God we know from faith.

Following Aquinas, you make it look as if the God of reason and the God of faith are the same God. But many would object to this, because they don't see how the God of reason and the God of faith can go together, or at least they notice some tension between the two.

Your point is clear and right. The God of the proofs of God's existence is the "God of philosophers and scholars"—which is not, as Blaise Pascal put it in his "Memorial," the "God of Abraham, the God of Isaac, and the God of Jacob."[88]

Nevertheless, I would maintain the God of reason and the God of faith are not as different as Pascal seems to suggest. They

---

[88] This text is taken from what is called the "Memorial," written on a piece of parchment and sewn into the lining of Pascal's coat.

should at least be compatible with each other. Aquinas is aware of the difference between the two. Yet he also realizes that the God of reason and the God of faith don't stand for two gods—they refer to one and the same God. That's why Aquinas considers it important to argue that the God of reason must be utterly unique and thus cannot be different from the God of faith. Expressed in more modern terminology, they may differ in sense, but their reference is the same—as the way the morning star and the evening star differ in sense but refer to the same entity, the planet Venus.[89]

Because God is utterly unique, there cannot be two Gods—one being the God of faith and another the God of reason. Aquinas proves this extensively in several questions after his five proofs of God's existence.[90] Once the utter uniqueness of "a god" has been shown, one can begin to use "God" as a proper name to refer to that one, utterly unique being, in which the God of reason and the God of faith come together in unison. So ultimately, the God of reason and the God of faith both refer to the same God—one God, not two.

Can't we just forget then about the God of reason and base our lives exclusively on the God of faith? The God of reason is just a very abstract God—Existence Itself, a Necessary Being, the First Cause, a Cosmic Designer, an Eternal Intellect. How could this abstract kind of God add anything substantial to the God of faith?

Don't underestimate the God of reason. St. Thomas makes it very clear that the God of reason—by whom everything else is created—"contains within Himself the whole perfection of being."[91]

---

[89] This distinction was introduced by the German philosopher and mathematician Gottlob Frege in 1892. He did so in his paper "On Sense and Reference," *Zeitschrift für Philosophie und philosophische Kritik*, vol. 100 (1892), 31.

[90] And also in book I of his *Summa contra Gentiles*.

[91] *Summa Theologica*, I, Q. 4, art. 2.

If the God of reason is Infinite Perfection,[92] then He must have infinite powers, specifically the powers of being all-powerful (omnipotent), all-present (omnipresent), all-knowing (omniscient), and all-good (omnibenevolent). These divine attributes are essential to God—without them, He would not be God. A god who is not all-powerful or all-present or all-knowing or all-good cannot be God. We will discuss this more extensively in the upcoming chapters.

The problem remains, though, that many people find it hard to see religious faith as a series of conclusions about God deduced from self-evident starting points or universal principles. The God they believe in is much richer than that. The power of reason pales in comparison with the power of faith. Why do you still want to promote reason when it comes to religion?

The Catholic Church has a longstanding tradition of promoting faith *and* reason. In his 1998 encyclical *Fides et Ratio* (Faith and Reason), Pope John Paul II warns us against not recognizing "the importance of rational knowledge and philosophical discourse for the understanding of faith, indeed for the very possibility of belief in God" (no. 55). In this encyclical, the pontiff also points out, "The Acts of the Apostles provides evidence that Christian proclamation was engaged from the very first with the philosophical currents of the time" (no. 36).

As a matter of fact, since early Christianity, reason and philosophy have been used to show us that there is a God and to demonstrate God's primary attributes, such as his power, knowledge, and presence. This way, reason lays the foundation for the God of faith and makes God's revelation to us through the Bible reasonable and also credible and therefore reliable. Reason is thus the common

---

[92] *Summa Theologica*, I, Q. 4, art. 1.

ground between believers and nonbelievers, which makes it an important apologetic tool in defending the Faith.

I have no problem with promoting reason on its own in the common issues of life, but is reason really necessary for our faith in God?

Yes, reason is necessary for religious faith too. If God were not "a God," then the God of faith wouldn't and couldn't be God to begin with. The God of reason is a "minimum" for the God of faith. Before we can believe what God has revealed to us in the Bible, for instance, we must first know that there is a God, that He is utterly unique and perfect, and that we have a soul capable of grasping philosophical truth. Belief in the God of reason — at least belief in God's existence — is a necessary condition for belief in the God of faith. It makes no sense, and is actually foolish, to believe in someone or something that doesn't exist.

No wonder, then, that Aquinas made an important distinction between the "articles of Christian faith" and the "preambles" to faith, the latter of which contain truths we can see with the natural light of reason and to which we must assent if we are to have faith at all.[93] Without assenting to the preambles, we could not possibly assent to the articles. Reason and philosophy can and must *prepare* the way to the God of faith. The *Catechism of the Catholic Church* acknowledges that the reasoned proofs of God's existence "can predispose one to faith and help one to see that faith is not opposed to reason" (no. 35). On the other hand, it is possible to believe that God exists without believing that God is a Triune God, a Trinity of Persons — but the reverse is impossible. Belief in the God of reason is not enough for belief in the God of faith — it's necessary but not sufficient.

Therefore, we must conclude that the God of faith is more than, but certainly not less than, the God of reason. This does

---

[93] *Summa Theologica*, II, Q. 5, art. 4.

not take anything away from the fact that God is much *more* than what mere human reasoning can tell us; we will discuss this more in chapters 11 and 12. At the same time, God is not *less* than what reason can reveal to us, for the power of reason comes from God too. That's why God must also be "a God."

In *Fides et Ratio*, Pope St. John Paul II said, "Faith and reason are like two wings on which the human spirit rises to the contemplation of truth; and God has placed in the human heart a desire to know the truth — in a word, to know himself — so that, by knowing and loving God, men and women may also come to the fullness of truth about themselves" (no. 1).

This takes us back to Aquinas: "We must believe that God exists, which is clear by reason."[94] What Aquinas means by this statement is that *reason* leads us, for instance, to assume a First Cause, whereas *faith* in God discovers that this concept refers to the God of our faith. The God of reason is not in conflict with the God of faith. And the God of faith doesn't take anything away from the God of reason. What remains standing is that the God of reason and the God of faith may differ in sense, but they both refer to the same God, who is utterly unique. That means that God is one, not two.

---

[94] Aquinas, *Compendium Theologiae*, 3.

7

Why God Is Invisible

Most ancient religions consider their gods as having physical bodies. In ancient Greek and Roman polytheism, different gods had their own temples, or one temple for all of them combined (the Pantheon). Pagans were able to see their gods in their temples anytime they wanted to. These gods were so visible and tangible that even the emperor, or the pharaoh, for that matter, could be seen as a god. But no matter what is being venerated—Jupiter, the Roman emperor, a golden calf, bulls, serpents, hawks, monkeys, or any other material entity—these gods are visible to the human eye. As one of the psalms puts it sneeringly, "They made a calf, bowed low before … the image of a grass-eating bull."[95]

How different is the God of reason! God being *invisible* is a rational conclusion. It follows from the proofs of God's existence. God, the First Cause of all material, secondary causes, cannot be material Himself, for that leads to infinite regress. The First Cause is not three-dimensional or four-dimensional or more-dimensional—no dimensions, period. Material things can move and change, but if the First Cause did, it would not be the First Cause. God as a Necessary Being cannot have a visible body, for bodies have parts, but a

[95] Ps. 106:19–20.

Necessary Being does not.[96] Everything that comes into being has parts that need to be combined, but Existence Itself does not come into being and has no need of combining parts. Instead, God is the noncomposite cause of all composite beings. Everything composite must ultimately depend on the existence of a noncomposite being. In short, unlike any contingent being, the God of reason is not contingent, not composite, not material, and not visible.

You claim that the God of reason doesn't have parts, yet you keep talking about God's attributes — being all-powerful, all-present, all-knowing, and all-good. Isn't that talking about God's parts, though? Aren't different attributes basically different parts of God?

Not at all. Those are merely different ways of talking or conceiving of one and the very same entity — God. We can focus on different "aspects" of this one unique reality, whether it is on God's power, on His knowledge, on His goodness, or whatever. These attributes differ in *sense*, but their *reference* is the same reality: God. God's infinity *is* his power, which *is* his goodness, which *is* his intellect, which *is* his will, and so on. They all come with the same unique reality, God. God is one and has no parts, otherwise He would not be God. We can talk about different attributes, but they all refer to the same undivided reality — God.

What we are basically doing here is using a form of abstraction. We have the ability to focus on one particular aspect or to take on one particular perspective, while leaving out of consideration other aspects or perspectives. That's how scientists can look at a human being from a physical, chemical, biological, or psychological perspective. But abstract perspectives are not concrete realities in themselves. A human being is a concrete reality, but in reality, there is no such thing as a physical human being or a chemical

---

[96] *Summa Theologica*, I, Q. 3, arts. 1–2.

human being. Telling things apart is not the same as setting them apart. Though we distinguish between them in *thought*, there is no distinction at all between them in *reality*. Think of this: the fact that we can distinguish a flame's heat from its light does not mean that we can separate the heat from the light.

OK, it's all about the same God, then. But I find it hard to believe in the existence of an invisible God. Almost everything we know has dimensions of length, width, and depth. We never bump into anything devoid of size, shape, and location. God doesn't have any of that. We cannot even imagine things like that. Doesn't that mean such a God cannot exist either?

Although everything visible is possible, it does not follow that something invisible is impossible. God may not be visible, but that does not mean such a God is impossible. The proofs of God's existence that we discussed earlier prove that God does and must exist as Existence Itself, the First Cause, a Necessary Being, the Cosmic Designer, an Eternal Intellect. All of these can be known by us through reason. These designations may be abstract, invisible, even unimaginable, but that doesn't make them nonexistent.

Perhaps an analogy may help us here. When visiting a university, we can walk around the campus and visit all its buildings, but we will never see a building for the university in and of itself. The university is made up of a collection of buildings, but the university itself is not a building visible among other buildings. Does that make the university nonexistent? Of course not. In an analogous way, God as a Necessary Being does not have a visible body as we do.[97] Yet God exists and makes our visible bodies possible, just as the university makes all the university buildings possible without being one of them.

---

[97] Ibid.

A Catholic Scientist Proves God Exists

I must admit that the proofs of God's existence are powerful, but they are not known to everyone. Because God is invisible, that might explain why people find it hard to believe that He exists. When we look through our telescopes or microscopes, we never see God, so we think we conclude from this that there is no God — on scientific grounds. How, then, could we ever know that God exists?

God's invisibility may make it harder to believe in Him, but I think many people use this as an alibi for *not* believing in Him. Their excuse during the Last Judgment might be what Bertrand Russell famously told us he would reply when facing God on judgment day: "Not enough evidence, God! Not enough evidence!"[98]

That's too easy a point to make. These people never gave God their time to study the evidence that points to Him or the reasoning that proves God's existence. Of course, one can be a nonbeliever in other things besides religion; many people are nonbelievers, for instance, when it comes to the existence of flying saucers or extraterrestrial life, because they never took the time to study such issues or to weigh the pros and cons. In other words, they are nonbelievers out of lack of interest. But when it comes to God, lack of interest is a much more serious case. We owe God our interest. But that assumes already that God exists, of course. God being invisible makes it certainly harder to become a believer.

We don't know why God decided to remain invisible. But the following analogies might be a bit of an answer. How could a Necessary Being be as visible as a contingent being? How could a First Cause be as visible as a secondary cause? How could the Designer of all there is be as visible as a human designer? How could an Eternal Intellect be as visible as a human intellect?

---

[98] Quoted in Wesley C. Salmon, "Religion and Science: A New Look at Hume's Dialogues," *Philosophical Studies* 33 (1978): 176.

Indeed, God is invisible, no matter how we turn it. But at the same time, we do have to add to this—in light of the God of faith, which will be discussed later—that God did become visible to us through His Son, Jesus Christ. As Jesus said to Philip, "Whoever has seen me has seen the Father."[99]

Though the God of reason is invisible, we often tend to talk about the God of faith as if He were visible. Religious people speak about God in very human terms. Doesn't that contradict what the proofs of God's existence tell them?

True, religious believers do speak about the God of faith in human terms. We have to understand such talk correctly, though. Let's not forget that St. Paul states emphatically, "At present we see indistinctly, as in a mirror, but then face to face."[100] St. Paul is very cautious here, although he does say "face to face," which sounds like very personal language. Why makes him dare to speak in those terms?

On the one hand, we need to stress that the God of reason is not merely something but *Someone*. God must be *more* than a human being, not less. If God were not a person, God would be *less* than we are, which would be contradictory to being God. St. Thomas Aquinas used to say, "You cannot give what you do not have."[101] Since God created us as persons, God must at least be a person Himself, but then a person with infinitely more power, infinitely more knowledge, infinitely more intellect, infinitely more love—having all of these characteristics to an infinite degree and with ultimate perfection. The bottom line is that God could never give us personhood if He didn't have personhood Himself.

---

[99] John 14:9.
[100] 1 Cor. 13:12.
[101] E.g. in *Summa Theologica*, I, Q. 4, art. 2.

On the other hand, we need to make clear also that God is not a person like you and me, for that would take away from God's perfection. When we call God a "person," we have to understand this term in an *analogous* way. God is not a person like you and me, and yet God is a person analogous to a person like you and me. When we say that God is able to "see," this is to be understood neither as a claim that God has eyes, as humans do, nor that God is incapable of seeing because He does not have such eyes.

What we mean instead is that God perceives in a manner analogous to the way humans perceive with their eyes. The psalmist could not have said it more clearly: "Does the one who shaped the ear not hear? The one who formed the eye not see?"[102] Indeed, how can God, who made the eye and the ear, not be able to see or hear? Remember, nothing can give what it doesn't have.

The Bible answers these rhetorical questions quite succinctly when it says that we were made after God's image and likeness.[103] Of course, God does not hear as we hear or see as we see. God alone is perfect—we are merely imperfect images of Him. That's why God sees and hears "infinitely" better than we do. The First Cause is analogous to, but in no way identical to, finite secondary causes such as you and me. God is not one of those secondary causes. So, when we talk about the invisible God in terms of visible features, we can do so only in an analogous way.

It always strikes me that Catholics in particular seem plainly to ignore that God is invisible. They portray God as a bearded father, for example. Isn't that in contradiction to one of their core beliefs?

Catholicism stands firmly in Judeo-Christian tradition. Judaism is very emphatic in stating that God is invisible. Even in the Temple

---

[102] Ps. 94:9.
[103] See Gen. 1:26.

they had in Jerusalem, the inner center didn't have an image of God, but only the Ark of the Covenant, which basically contained no more than the Ten Commandments. The First Commandment is very definite about an invisible God: "You shall not make for yourself an idol or a likeness of anything in the heavens above or on the earth below or in the waters beneath the earth; you shall not bow down before them or serve them."[104] Jews even go as far as to refrain from pronouncing God's *name*, Yahweh.

Why did the Catholic Church not enforce this commandment more strictly? First of all, the First Commandment is about using idols, about objects of worship, not about using images or icons of God. Images or icons are different from idols. The Old Testament tells us that God did, in fact, reveal Himself under visible forms—for instance, in the book of Daniel: "As I watched, thrones were set up and the Ancient of Days took his throne. His clothing was white as snow, the hair on his head like pure wool; his throne was flames of fire, with wheels of burning fire."[105]

Most of us make depictions of God the Father with forms like these. They are merely icons, not idols.

But, more importantly, in Jesus Christ God showed mankind an "icon" of Himself. St. Paul said, "He is the image [*ikon* in Greek] of the invisible God, the firstborn of all creation."[106] Christ is the tangible, divine "icon" of the unseen, infinite God. Jesus changed a "lofty" God into a close-by Father and told us to pray to Him as a Father.

What God forbids is not using images but "worshiping them as gods"—replacing the invisible God with idols we carve, depending on them, bowing to them, and ascribing to them attributes

---

[104] Exod. 20:4–5.
[105] Dan. 7:9.
[106] Col. 1:15.

that belong to God alone. Even if we have an image of God—for instance, the image of God the Father, with white hair, sitting on his throne in Heaven—this image does not replace God but only "pictures" God. In a sense, pictures are similar to words. When we speak about God in language, we describe what we imagine God is like; in a similar way, when we visualize God in art or illustrations, we express how we picture God.

The Fourth General Council of Constantinople (869–870) makes the same comparison: "What speech conveys in words, pictures announce and bring out in colors." In this context, it is worthwhile quoting St. Germanus, the Patriarch of Constantinople, who battled the iconoclasts of his time: "Pictures are history in figure.... We do not worship the colors laid on the wood."[107]

---

[107] In his letter to Thomas of Claudiopoli.

8

Why God Is All-Perfect

The wording *"all-perfect"* may sound like a duplication, yet I use that word to make sure the term *perfect* is not understood in the way things such as cars are perfect for the purpose they were made for. God is perfect, not in a relative sense but in an absolute sense: all-perfect. When we say that God is all-perfect, we mean that He is unlimited in every kind of perfection, or that every conceivable perfection belongs to God in the highest conceivable way. As St. Thomas Aquinas put it, God "contains within himself the whole perfection of being."[108] God's perfection follows directly from what God *is*.

Being all-perfect, God has no deficiencies. His perfection is grounded in the fact that He is Existence Itself and thus encompasses all contingent beings. The First Vatican Council explicitly taught the dogma of the perfection of God.[109] Additionally, the doctrine is based on Jesus' command, "Be perfect, just as your heavenly Father is perfect."[110]

If it is true that God is all-perfect while the world is not, then the distance between the world and God must be infinite. Doesn't

[108] *Summa Theologica*, I, Q. 4, art. 2.
[109] *Dei Filius*, I: DS 3002.
[110] Matt. 5:48.

this cause an infinite gap between God and us? How can finite beings ever cross an infinite gap?

There are two extreme positions in answering this question. We either claim that God is like all other entities—which is obviously false, for God cannot possibly be like all other entities, as we found out in the proofs of God's existence—or we must admit that we cannot possibly know anything about God—which is also untrue, for if we don't know anything about Him, we cannot even know or say that God exists.

Unfortunately, many philosophers have taken the latter position. They seem to make it their profession to doubt everything—they are the professional skeptics among us. They often go as far as doubting whether you and I exist, or whether there can be any truth at all. Amusingly enough, scientists sometimes joke about their work with warnings such as these: "Don't touch anything in a physics lab"; "Don't taste anything in a chemistry lab"; "Don't smell anything in a biology lab." But perhaps we should add also this warning: "Don't trust anything you hear in a philosophy department." We have indeed much reason not to trust those skeptical philosophical views.

Some of those professional skeptic philosophers consider themselves followers of the German philosopher Immanuel Kant, who placed an insurmountable distance between God and the world, which made God completely inaccessible to human experience. They declared God so "totally other" that we cannot really know anything about Him. They declared Him "unknowable."

A first response to these skeptics should be that they seem to know at least one thing about the unknown God—that God is unknowable! Do we not know more than just this? Can we not make the supposedly "unknown God" a bit more known? As a matter of fact, there is nothing wrong with thinking that God can be known (to a certain extent) based on what we see and experience around us, combined with the power of reason.

Apparently both extreme positions are indefensible. Once we accept that we can at least say something about God, we have two options left. The first one is that our talk about God would be considered so open-ended that it becomes completely inadequate and empty. If that were true, then religious faith would come very close to "blind" faith.

The other option is that we can speak about God only in purely negative terms—"God is *not* this" and "God is *not* that." The problem with this latter approach is that one cannot merely say that God is not this and not that without saying something positive. A negative definition of God proceeds by elimination—it can begin, it can go on indefinitely, but it can never do its job. The best we could ever say this way is "O God, if there is a God, save my soul, if I have one."

Another, and arguably better, way of talking about God is this: God is *transcendent* to the world and to all that is in it. Most religions speak about God in terms of *transcendence*. This means we are capable and allowed to speak about the transcendent God, but only in analogous terms.

The question remains, though, as to how the finite human mind could ever reach out to a transcendent God? Doesn't God's transcendence prevent that?

First of all, transcendence is something everyone understands intuitively. Human beings do not possess a complete grasp of themselves—they are a mystery even to themselves. When I say, "I am only human," I am not comparing myself with something "below" me—such as a cat, a dog, or a chimp; I am comparing myself with Someone who is "above" me and transcends me. When I call myself "only human," I am comparing myself with Someone who does not have the limitations and deficiencies I experience. In some mysterious way, I am reaching out into the realm of the Absolute, far beyond myself. It's the realm of the all-perfect God.

How can this be? Apparently, the "finite" human mind is able to catch no more than a "glimpse" of the Infinite. To use a poor analogy: if I can count from one to ten, or from one to a thousand, then I can count also from one to infinity, at least in theory. This faculty of the human mind for the infinite allows us to transcend ourselves and reach out to the transcendence of an Infinite Being far beyond ourselves. Belief in the transcendent is at the heart of all religions. Of course, I cannot transcend myself on my own; that is possible only because I was made in the image of God, and I perceive more than myself when I perceive myself completely.

Second, transcendence, too, is often misunderstood as an uncrossable "infinite gap" between the transcendent, infinite nature of God and the finitude of human beings—specifically with regard to reason. This gap is supposed to keep us from being able to reason effectively on religious matters. But again, that's a one-sided view. C.S. Lewis tried to show this when he noted that "if there was a controlling power outside the Universe, it could not show itself to us as one of the facts inside the Universe—no more than the architect of a house could actually be a wall or staircase or fireplace in that house."[111]

And yet we have some access to the Architect of this universe. Every analogy is limited, but just as architects and builders are not parts of their buildings yet are somehow "part" of every part of them, so is God transcendent (that is, not a physical part of what He created) yet immanent (that is, actively involved with each and every part of it at each and every moment). God transcends what He created and yet is fully present to it—all-present, that is. We

---

[111] C.S. Lewis, Mere Christianity, bk. 1, chap. 4.

are just God's creatures—not "next" to God but "under" God, not "outside" God but "in" God. "Outside" God, there is nothing—no space and no time. As St. Hilary of Poitiers said about God, "He is outside of all things and within all things."[112] God is like the architect who is not part of the building, yet the building is somehow part of him.

God's transcendence—being infinitely far above all we know —should always be in balance with God's immanence—His being closely and intimately involved with His creatures. Jews and Christians always try to balance carefully transcendence and immanence so God does not become either too human (like deities such as Jupiter) or too distant (like a "totally other" God). God is part of everything (immanent) without being a physical part of it (transcendent). God's closeness and His superiority are two sides of the same coin. Whoever denies one of these sides detracts from the God of reason and also from the God of the Judeo-Christian faith.

It is God's immanence in the world that gives us signs of the Maker of this world. St. Paul said about God that "his invisible nature, namely, his eternal power and deity, has been clearly perceived in the things that have been made."[113] St. Paul made it clear that we can see the invisible God *through* what is visible—God's immanence. What is visible points to the invisible God. This insight, however, is based more on induction than on deduction. It doesn't have the rigor of deductive reasoning. We are dealing here with pointers rather than conclusive evidence.

As a matter of fact, the previous chapters showed us we have additionally very compelling, deductive reasons to say that God is all-perfect. Therefore, God's transcendence doesn't prevent us from saying something positive about God after all. Sure, we do not

---

[112] Hilary of Poitiers, *On the Trinity*, 2, 6.
[113] Rom. 1:20.

know everything about God, but we know at least certain things: that God exists and has certain divine attributes.[114] In other words, reason allows us to know *that* God is and even a bit of *what* or *who* He is. The rest has to come from faith.

---

[114] Thomas Aquinas, *Summa contra Gentiles*, I, 30.

9

Why God Is All-Powerful

God's being *all-powerful* is another rational conclusion that follows
from the proofs of God's existence. *Power* is the "capacity to act
or to make." If God is the First Cause of the world, then obviously there is power in Him, for as St. Thomas says, "all operation
proceeds from power."[115] But God must also be infinite in power,
according to Aquinas, because "the more actual a thing is the more
it abounds in power."[116] That from which all power derives, without
having anything outside its range of power, must be all-powerful
or omnipotent.

It is important to realize, though, that the power of God's omnipotence does not merely exceed other powers; it *transcends* other
powers. God's omnipotence is not the sum total of all human powers. God's power is not like our powers raised to the power of a
zillion. Instead, God's power is the source and origin that all other
causal powers depend on and derive from; those other powers are
not "next" to God but are "under" God. They would not have any
power if God did not give them some power. Somehow they share
in God's power.

---

[115] *Summa Theologica*, I, Q. 25, art. 1.
[116] *Summa Theologica*, I, Q. 25, art. 2.

A Catholic Scientist Proves God Exists

God's being all-powerful is a nice thought, but it has made some people reject God as a dictator.

Yes, some atheists take this position. Some think that believing in God is believing in a sort of "benevolent dictatorship," as they call it.[117] Probably the best known representative of a similar view is the French philosopher and atheist Jean-Paul Sartre.[118] He took it that an almighty God does not leave any room for free human beings, and free human beings do not leave room for an almighty God. In Sartre's eyes, God and man are in a perpetual power battle. Sartre opted for human freedom over divine omnipotence and thus became an atheist (until just before he died). Does he have a point here?

I would say he does not — for at least three reasons. The first reason is that Sartre's dilemma puts God and man at the same level, whereas they certainly are at very different levels. God is not one person among other persons, just as He is not one cause among other causes. There is no competition between God and His creatures. Therefore, submitting ourselves to God, the Maker of Heaven and Earth, is not like submitting ourselves to a dictator, who is just another person in our midst.

On the contrary, the more we become like God, the more we become like ourselves, since we were made in God's image. Yet we tend to think that our freedom is compromised by God's omnipotence, so we take it as a threat to our freedom and think we should rebel against it — as if God's power and our power compete with each other. We think we are free only when we are free of God. But

---

[117] The late British atheist Christopher Hitchens is one of them. See his 2007 book, *God Is Not Great*.

[118] Jean-Paul Sartre asserts in *Being and Nothingness* that man is a creature haunted by a vision of "completion," called by Sartre the *ens causa sui*, which religions identify as God.

the opposite is the case: following God's will is an opportunity for us to become more like our deepest self. That's quite a different take.

My second reason for rejecting Sartre's dilemma is that God's omnipotence does not take my freedom away. God's power is not a blind brute force like the forces we are familiar with, but a power far beyond our comprehension. As a matter of fact, God decided in His omnipotence to make human beings participants and coworkers in His creation, in accordance with His image. He chose in His own freedom to endow us with freedom, too, as a reflection of His freedom. Consequently, creation did not spring forth complete from the hands of the Creator. God gave us also the dignity of acting on our own, and thus of cooperating in the accomplishment of His plan, by enabling us to be intelligent and free causes of our own in order to complete the work of creation. More on this later.

My third reason for rejecting Sartre's dilemma is that human freedom would remain hanging in the air if there were no God. We have nothing to base it on, other than God. It certainly cannot be based on genetics. Theoretically, there could indeed be a gene in our DNA that allows us to make choices, but if this gene, or any additional genes, would also determine the outcome of these choices, then we could not really make free choices and would basically have lost the free will we thought we had. If we want to claim human freedom, we need someone from whom this freedom derives—a Creator God. So it is not a matter of God *or* freedom but of God *and* freedom.

In other words, there can definitely be human freedom under an all-powerful God. Dictators may take human freedom away, but God made us in His image, and thus He created us not as marionettes but as beings endowed with freedom as well. God gave us freedom in His infinite, selfless love. Because we are rational beings made in God's image, we have been created with free will. In other words, the right to choose is ours. God lets the actors on the world stage be free actors, who may not act the way the Author of the play would like them to act. We are master over our choices, despite the fact that they will be misused over and over again in error and sin.

So, humans do not have to be powerless if there is an almighty God. Sartre was wrong on that issue and created a false dilemma: no longer do we have to question God's omnipotence, or even His existence, because there is human freedom. My freedom to make my decisions does not mean that this freedom takes away from God's sovereignty and omnipotence.

Even if it's true that God's being all-powerful does not amount to a dictatorship, it still does take away from our freedom to act, I would say. If God is all-powerful, the free will we think we have must be an illusion. Whatever I think I am doing on my own is actually God's doing.

Yes, it does seem as if there is another conflict here between human freedom and God's omnipotence. Let me make clear first that the existence of our free will is not an illusion. One of the pioneers in neurosurgery, Wilder Penfield, made a compelling case for our free will when he asked one of his patients, during open-brain surgery, to try to resist the movement of the patient's left arm that Penfield was about to make move by stimulating the motor cortex in the right hemisphere of the patient's brain. The patient grabbed his left arm with his right hand, attempting to restrict the movement that was to be induced by a surgical stimulation of the right brain.

As Penfield described this: "Behind the brain action of one hemisphere was the patient's mind. Behind the action of the other hemisphere was the electrode."[119] In other words, one action had a physical, neural cause, whereas the other action was ruled by a

---

[119] Wilder Penfield, in the Control of the Mind Symposium, held at the University of California Medical Center, San Francisco,

mental cause, the patient's free will. Therefore, Penfield concluded, the physical cause and the mental cause had different origins and were of different natures.

The seeming conflict between human freedom and God's omnipotence is of an entirely different nature, though. St. Thomas Aquinas explained this apparent conflict between God's activity as Creator and ours as free creatures by using his powerful distinction between First Cause and secondary causes.

Without Aquinas's distinction, we are prone to think of God's activities as some kind of secondary cause, situated at the same level as our free decisions. We think, for instance, that God creatively wills that I decide to do something, and His willing then causes me to make a decision—as if His will were a secondary cause. In this scenario, God's creative fiat would be an event independent of my decision and would indeed rob me of my autonomy if that were true. I would be no longer a free agent but a puppet manipulated by God.

But this interpretation is flawed. In contrast, it needs to be stated that God, as a First Cause, works in such a way that we are not acted upon as if God were a secondary cause; instead, we exercise our free will as a secondary cause of our decisions and actions. God is not the direct cause of my decisions—I am—but He is the *indirect* cause that lets me be the cause of my own decisions. Seen in this light, our human freedom need not be in conflict with an all-powerful God at all. God is the complete and First Cause of a free act, whereas the human agent is its complete secondary cause. In other words, my freedom to make my own decisions does not mean that this freedom takes anything away from God's sovereignty and omnipotence.

1961. Quoted in Arthur Koestler, *Ghost in the Machine* (London: Hutchison Publishing Group, 1967), 203–204.

A Catholic Scientist Proves God Exists

To use an analogy, God makes His creatures movable by nature so that they can move themselves. In a similar way, God gives us the power to make decisions. Thinking that God and man are in a power battle basically amounts to thinking that if God does something, then nothing or no one else is doing it, and if something else is doing it, then God is not the cause of it. It is God's power that is the source and origin of all human power.

Still, God's being all-powerful cannot be taken literally. If it were, then God would have the power to make a stone so heavy that even He could not even lift it. That seems contradictory to me.

St. Thomas denies strongly that there is a contradiction here. He is very definite in defending that God cannot act against reason. When something is against reason, God cannot create it.[120] Aquinas is very adamant on this issue because God is reason Himself, so He cannot act against His own nature by doing what is contradictory. God is absolutely free, but His freedom is not arbitrary, so He cannot go against what is true and right. How do we know this? Because our own power of reason is rooted in creation and thus participates in God's power of reason.

As a consequence, God's being all-powerful does not mean that He is able to do what is logically contradictory.[121] Aquinas gives many examples of what God is not able to do: God cannot create square circles; He cannot create triangles with four sides; He cannot make anyone blind and not blind at the same time; He cannot declare true what is false; He cannot declare right what is wrong; He cannot undo something that happened in the past; and the list goes on and on.[122] To use your example: God does not

---

[120] Thomas Aquinas, *De Aeternitate Mundi*, I, 4.
[121] Thomas Aquinas, *Summa Theologica*, I, Q. 25, art. 3.
[122] Thomas Aquinas, *Summa contra Gentiles*, II, 25.

even have the power to make a stone so heavy that He Himself cannot lift it—that would indeed be contradictory, and therefore against reason, and therefore impossible. Aquinas concludes from this: "Hence it is better to say that such things cannot be done, than that God cannot do them."[123]

This also explains why God never acts against the laws of nature. These laws are contingent—they could have been different or even nonexistent—and therefore are not self-explanatory. Yet they happen to be the way they are. They are this way because they were implemented by God. Without the law of gravity, for instance, we would fall off the earth, and the earth and the other planets would not be able to remain in stable orbits around the sun. It is thanks to the law of gravity that God made a universe in which small objects are attracted to larger ones. It would be contradictory for God to go against His own laws.

Since gravity is a secondary cause, God made a universe in which He does not have to be the direct cause of every stone falling to the ground. Thanks to the laws of nature, we are able to know what the outcome is of certain contingent events such as stones falling to the earth. We do not have to wonder about God's will every time a stone falls to the ground, even if it strikes us on the head. God has given us a secondary cause—the force of gravity—which is the direct cause of each stone's earthly plummet. This way, we live in a world that we can trust and count on because it follows the God-given laws of nature. Knowing this helps us make decisions that are not only free but also based on sound, reliable predictions.

The way you describe things makes it look as if God has delegated His power to the laws of nature. If that is true, then God seems to be at the mercy of His own laws.

---

[123] Thomas Aquinas, *Summa Theologica*, I, Q. 25, art. 3.

A Catholic Scientist Proves God Exists

It may seem that way, but God is not at the mercy of His own laws of nature. Let me explain this with a simple analogy — imperfect like all analogies yet very telling. When watching a game on the golf course or on the pool table, we see balls following precisely determined courses of cause and effect; they follow physical laws and are subject to a cause-and-effect mechanism. Yet there is one element that does not seem to fit in this cascade of causes and effects — the players of the game. Although we have here a cascade of physical causes and effects, there is much more going on during these games: the players have a specific intention in mind, which is the power of human freedom. Their free decisions elude and transcend the laws of nature. They don't go *against* the laws of nature, nor do they change them — yet these players do go *beyond* them by using and steering them according to their free decisions. People who are unable to look beyond these physical laws and causes are completely missing out on what the game is all about.

The point is this: If we are able to steer the laws of nature ourselves, why would God not be able to do the same — in an analogous way, of course? God is actively present in this world, not by going against the laws of nature and its secondary causes, or by supplementing or replacing them — such things cannot be done, Aquinas would say — but by letting them be the way they are and yet going beyond them by steering them in a certain direction without overstepping the autonomy of secondary causes. Of course, that is not all there is to it, but perhaps this analogy opens the door for us to understand better what God's omnipotence is like in this universe. Nature is ruled by its own laws of nature, yet it is not autonomous, for God upholds and directs the world by His creative, steering, sustaining activity.

I think the worst problem of God's being all-powerful is something you might like to ignore: the existence of evil in the world. How can evil ever be reconciled with an all-powerful God?

In one of the next chapters I will delve into this issue more exten-sively. For now, let me say just this. It is clear that the possibility of evil in this world is a consequence of human freedom too. How could God ever give us freedom without accepting its consequences up to the point of our freely choosing the wrong outcome, away from God and in opposition to His divine will? If our world were different, fully preordained by God, our brains would automatically refuse to enforce what our free minds would like to do. But that would be a travesty of human freedom: "a toy world which only moves when He pulls the strings," in the words of C. S. Lewis.[124] I promise to get back to this later in the book when we talk about God's being all-good.

For now, we may conclude that the almighty God gave second-ary causes their own power; without God, they would have no power. He also gave human beings decision-making power; without God, they would have no power at all. God even allows them to unjustly call Him a "dictator."

[124] Lewis, *Mere Christianity*, bk. 2, chap. 3.

10

Why God Is All-Present

God's being *all-present* is another rational conclusion that follows from the proofs of God's existence. There are at least two dimensions to being all-present: timelessness and spacelessness. As to being timeless: God must exist outside of time, for if He existed within time, He would constantly be adding time to His existence, which means change; but a god who changes is no longer God. As to being spaceless: God must exist outside of space, for if God were three- or four-dimensional, He would be located somewhere in space, which would make Him lose His status of being God.

Therefore, God does not exist at a specific place or at a specific time. Instead, God is all-present, being outside of time and outside of space. God is all-present in both space and time, being both timeless and spaceless. This explains, for instance, why God cannot be put to the test in a laboratory, as we mentioned earlier, for it would be in contradiction to His omnipresence—His being timeless and spaceless. How can this be? Let's discuss the questions of time and space separately.

Seen in terms of *space*, God is everywhere. He is not located in any specific spot; therefore, any attempt to identify Him with a particular location, or even with the sum of all locations, necessarily fails. God transcends all spatiality. That's why it may seem that God is nowhere, because He is "everywhere" in terms of space. In other words, because God is everywhere, it only appears as if He is "nowhere" in particular. That's why God cannot be "seen" through telescopes or microscopes, because that would limit His being all-present.

As it has metaphorically and paradoxically been expressed a few times, "God's center is everywhere, His circumference nowhere."[125] That God is truly present in every place or thing or situation follows from the fact that God is the ultimate cause and ground of all reality. God is not somewhere "in" space—He is the one who created space. Without God, there couldn't even be any space.

Not only is God all-present in terms of space, but He is also all-present in terms of *time*. This is often expressed as God's being eternal, which means He has neither a beginning nor an end. Consequently, God is at the beginning of everything and will still be at the end of everything. There is no past or future for God—only an eternal present. With time being a measure of finite existence, the Infinite One cannot be bound by it. God's existence is timeless. Past and future do not apply to God, who, in the words of St. Augustine, lives in "the sublimity of an ever-present eternity."[126] In that sense, God is the Alpha and the Omega by enclosing everything in between. God is not somewhere "in" time; He is the very Creator of time—without God, there couldn't be any time.

Earlier we said that God is invisible. Now we can add to it that God is also all-present. The One who is outside of time and space must be invisible. Yet God produces effects that are within time

---

[125] This thought was used by Cardinal Nicolaus Cusanus and Pascal, among others.
[126] Augustine, *Confessions*, bk. 11, chap. 13.

and space in the universe. This makes Him seem to be nowhere in the chain of worldly events, because He is everywhere all the time, being the origin of all that comes into being anywhere. And that's why science, for example, is "blind" to God's existence.

Aren't space and time part of everything we are familiar with in this world? How could we ever think or talk about anything outside of space and time? That seems like a mind-boggling challenge to me.

I grant you that all contingent beings are indeed spatiotemporal—always existing at a specific place and at a specific time. Space and time are definitely part of the created, material world. Albert Einstein has already shown us that both time and space are as much a part of the physical world as matter and energy are. In point of fact, time can be manipulated in the laboratory. The presence of mass (or more generally energy) causes space-time to curve. Spatiotemporal categories are part and parcel of this universe. We didn't really need Einstein to come to this realization.

Space and time, however, come only with contingent beings. It does not follow from this that God must have space and time too. Unlike contingent entities, which must come into being, God is not a spatiotemporal entity but is Existence Itself. Since both space and time are measures of the finite, God's infinity must transcend all temporal and spatial limitations. This may sound highly abstract, but it can be boiled down to this: God is an infinite power beyond the spatiotemporal dimensions of our universe.

In other words, God, who is the origin of all that exists in the world, did not create the world *in* time, but *with* time—or having time in it. Neither did God create the world in space, but He created space in the world—or a world full of space. From out of His very being, God gives being to the whole world and all things in it, including time and space. They could not exist without a timeless and spaceless God.

A Catholic Scientist Proves God Exists

I have news for you. If God has something to do with the Big Bang, as many people believe He does, then God must somehow be spatiotemporal too, for He started the universe at the beginning of time, and He made the universe stretch its space.

The confusion behind what you say might be that you identify the Big Bang with creation. But those two concepts are certainly not the same. The Big Bang was indeed the beginning of space and time. The universe started to stretch its space at a certain beginning in time — some fourteen billion years ago. Since space-time is something that has not always existed, it makes no sense to talk about a "time" when there was no time and space yet.

What about creation, then? Why can creation not happen at a certain time? The reason is simple. Creation cannot follow a timeline, as time itself has to be created to begin with. Time cannot exist before there was time, because there can be no time "before" there was time. As shown earlier, time is part of the physical world as much as space is. Time is part of something that must exist already, the physical world; otherwise there wouldn't be any time. Even St. Augustine knew this long ago, when he said, "There can be no time without creation."[127] He did not need science to come to this conclusion. Science would not even be able to come to this conclusion.

In other words, creation is not an event at all, and therefore it eludes science. Creation is not a "one-time deal," but it answers questions about where this universe ultimately comes from, how it came into being, and how it stays in existence. The answer to these questions is that the universe, and everything in it, does not come from the Big Bang but may have started with the Big Bang. The universe is ultimately something that finds its origin and foundation

[127] Ibid., bk. 11, chap. 30.

somewhere else. Without creation, there could not be anything—no Big Bang, no gravity, no evolution, not even a timeline. Creation sets the "stage" for all of these things and keeps all of them in existence. That's basically what all proofs of God's existence are about. Because of this, creation shares in the timelessness of the Creator. It is not something that happened long ago in time, and neither is the Creator someone who did something in the distant past, for the Creator does something at all times—by permanently keeping our contingent world in existence. Whereas the universe may have a beginning and a timeline, creation itself does not have a beginning or a timeline; creation actually makes the beginning of the universe and its timeline possible. William E. Carroll is right when he stresses that we should never confuse temporal *beginnings* with eternal *origins*. In his own words: "We do not get closer to creation by getting closer to the Big Bang."[128] To put it differently, we do not get closer to our origin by getting closer to the beginning of this universe.

In summary, creation is an act of the eternal, all-present God outside of time, but it produces effects that are within time. As Stephen Barr puts it, "The divine act of creation 'precedes' the existence of the universe only in a causal sense, not in a temporal sense. It did not happen in a 'time before' the universe came to be."[129] Edward Feser adds, "The whole idea that God exists timelessly is precisely that he does not exist at some point *in* time, but rather *outside of* time altogether."[130]

Perhaps we can use an analogy to make this insight more tangible. God's omnipresence makes Him the "soul" who pervades

---

[128] William Carroll, "The Genesis Machine: Physics and Creation," *Modern Age* (Winter/Spring 2011).

[129] Barr, *Modern Physics*, 263.

[130] Feser, *Five Proofs*, 202.

the universe. Just as my soul is part of everything I do with my body — but without being a bodily part — so is God part of every-thing that takes place in the space and time of the universe — but without becoming a cosmic part of it located somewhere in space and time. The God who is transcendent is immanent in what He created. As the soul is "all-present" in the body, so is God all-present in His creation. But that's where the analogy ends, of course.

11

Why God Is All-Knowing

God's being *all-knowing* is another rational conclusion that follows from the proofs of God's existence. If God didn't have perfect knowledge, then something would be missing in His knowledge, which would call for further explanations. If God had deficiencies in His knowledge, then His knowledge would be incomplete, and He would not be God.

God's omniscience entails the most perfect knowledge of all things. In the first place, God knows and comprehends Himself fully and adequately, and second, God knows all created objects and comprehends their finite and contingent mode of being. It is on Himself alone that God depends for His knowledge. If God were in any way dependent on creatures for His knowledge of created objects, that would destroy His infinite perfection. Being all-knowing is essential to being God.

We see here, as we saw with God's being all-powerful, that God's being all-knowing does not mean that He can know the unthinkable—that would be contradictory. For example, God could never think of the number eleven as an even number or of a circle as a square. Omniscience does not extend to things that cannot be but only to things able to be.

A Catholic Scientist Proves God Exists

Also, God's omniscience is not the sum total of all human knowledge, as it surpasses it in an infinite way—just as omnipotence is not the sum total of all human powers. In the argument from universals, we found out that concepts, such as *whiteness*, including all the statements that contain such concepts, must exist in God's Eternal Intellect. Since God has knowledge of *all* possible, actual, or necessary concepts and statements, God must have knowledge of everything—which is the essence of His being all-knowing or omniscient.

I find it easier to accept that God knows the past and the present, but God's knowledge of the future looks to me like a form of foreknowledge. If God does indeed know what the future will bring, then we end up with complete determinism that determines the future—with clockwork precision, one might add.

If you are right, then we have a serious problem. Complete determinism is something we naturally tend to reject. In science, however, some form of determinism may make sense. It allows us to predict tomorrow's weather, as today's weather depends on what happened yesterday, and tomorrow's weather will depend on today's. As the French astronomer Pierre-Simon Laplace put it more extensively, "We may regard the present state of the universe as the effect of its past and as the cause of its future."[131]

In other words, science seems to indicate that the future is determined by the past. Even Hollywood dealt with this issue in the 1944 movie *It Happened Tomorrow*. It portrayed someone mysteriously receiving the newspaper of the next day. This interesting feature enabled the man to be prepared for what was ahead of him—including an assault on his life, which he tried to escape, albeit in

---

[131] Pierre-Simon Laplace (1749–1827), *A Philosophical Essay on Probabilities*, 6th ed., trans. F. W. Truscott and F. L. Emory (New York: Dover Publications, 1951), 4.

vain, of course. It is still very much debated whether determinism can go that far. We all know how hard it is make predictions! Let's leave it at that.

On the other hand, when it comes to human freedom, determinism is widely rejected, because it goes against common sense and against our strong intuitive feeling that we are free beings with decision-making power. The late Nobel laureate and physicist Arthur Compton, who discovered the Compton effect,[132] expressed his rejection of determinism very strongly when he said, "If the laws of physics ever should come to contradict my conviction that I can move my little finger at will, then all the laws of physics should be revised and reformulated."[133] Indeed, having a free will is something very dear to us. St. Thomas Aquinas defines it as follows: "Free-will is the cause of its own movement because by his free-will man moves himself to act."[134] If that's true, and if my future decisions are not fully predetermined, then the future cannot be fully determined. This causes trouble for the doctrine of determinism.

As a matter of fact, determinism in its all-inclusive version causes several problems. First of all, it would lead us into a vicious circle of infinite regress—with no way to get out. Here is why. Determinism posits an infinite regress of causes—cause n is caused by cause $n\text{-}1$, and so on backward in time or space. Let's say the particles studied in particle physics are predetermined, but then the question arises: "Determined by what?" By smaller particles? What, then, determines these particles? Even smaller and smaller particles? The chain of causes would have to regress infinitely—which is

---

[132] The Compton effect is the increase in the wavelengths of X-rays and gamma rays when they collide with and are scattered from loosely bound electrons in matter.

[133] Arthur Compton, *Man's Destiny in Eternity: A Book from a Symposium*, The Garvin Lectures (Boston: Beacon Press, 1949).

[134] *Summa Theologica*, I, Q. 83, art. 1.

in itself problematic because an infinite number of things cannot exist in a material world.

Second, universal determinism creates a loop that makes for a logical paradox caused by the problem of self-reference. Self-reference is used to denote a statement that refers to itself. The most famous example of a self-referential sentence is the so-called liar sentence: "This sentence is not true." If we assume the sentence to be true, then what it states must be false. If, on the other hand, we assume it to be false, then what it states is true. In either case we are led to a contradiction.

Well, trying to convince someone of the truth of universal determinism smacks also of self-refutation. How could this claim change someone's mind if everything is fully predetermined anyway? Ironically, people who defend universal determinism are willing to spend their entire careers forcing everyone else to choose their deterministic conviction that human beings cannot choose. Chesterton said once in his characteristically pungent way that, if the world is determined, it makes no sense to say "thank you" to the waiter for bringing the mustard.[135] We can always say to dedicated believers in the doctrine of determinism, "Predict my behavior, and I will do otherwise." Of course, the real fanatics may counter that this would be predictable as well had they been given "enough" information.

Other dedicated believers in determinism have argued that determinism cannot refer to itself. They received some support from the British philosopher Bertrand Russell, who tried to argue that nothing can logically refer to *itself*—which is called his theory of logical types.[136] Thus, determinism could be one of those cases where

---

[135] G. K. Chesterton, *Orthodoxy*, chap. 2.
[136] Alfred North Whitehead and Bertrand Russell, *Principia Mathematica*, rev. ed. (Cambridge, UK: Cambridge University Press, 1925), 37.

"self-reference" is not allowed. Russell's argument in itself, however, is also a statement that defeats itself, since, in presenting the theory of types, Russell can hardly avoid referring to the theory of types, which is doing something that he tells us can't be done or shouldn't be done. His rule against self-reference can hardly be true. Just take the word *word* and ask whether it refers to itself. Of course it does—the word *word* is a word. So, excluding or forbidding self-reference is not a viable way to follow, not even to defend determinism.

Third, universal determinism acts as a boomerang that hits the one who launched it. If genes, for instance, really determine everything in one's life, they would also determine one's decision to believe or not to believe the claim that genes determine everything in life. The key problem is that we are dealing here with *beliefs*, and beliefs are not material entities like genes—for unlike genes, they can be true or false. If I believe that genes determine everything, I have no reason to suppose my belief is true, and hence I have no reason for supposing that genes determine everything. This is a "boomerang theory" in *optima forma*—it defeats itself, for once we consider it to be true, it becomes false. In a similar way, universal determinism defeats itself.

You still haven't addressed my suggestion that God's omniscience necessarily leads to determinism. If God has knowledge of the future, then the future is predetermined.

That is not necessarily true. Perhaps a simple analogy will help explain why God's omniscience doesn't imply determinism.

Picture yourself watching a video of certain events in your past. You may get the impression that these actions were in fact predetermined and preordained, and yet you know that most of them were freely decided upon when they were taking place. Your knowing and seeing that they happened does not mean they were predetermined.

Obviously, we cannot apply this analogy directly to God, as if God were watching a video of what is taking place on earth in the past, present, and future. What is definitely wrong in this analogy is that it might suggest that God's foreknowledge is the result of His watching what happens as history unfolds. God doesn't need to watch how things happen in order to know that they happen. God knows everything by virtue of being the First Cause of everything that happens. As Edward Feser puts it, "His knowledge of the world is a consequence of his self-knowledge."[137]

Another reason why God's knowledge of the future does not imply determinism is Aquinas's distinction between the First Cause and secondary causes. We think, for instance, that God knows ahead of time that I will decide to do something, and God's foreknowledge then causes me to make that decision—as if God's foreknowledge were a secondary cause. That could indeed lead to determinism. In this scenario, God's knowing would be independent of my willing, which would indeed rob me of my autonomy. I would no longer be a free agent, but merely a puppet manipulated by an all-knowing God. That's what Aquinas attacks as a mistaken belief.

I still maintain that if God foreknows my decisions, my decisions don't count. So it doesn't make sense for me to decide on marrying someone, or on choosing a career, or on believing in God, or on anything else. God's omniscience makes the past and the present and the future all together one long, accomplished fact. Doesn't this suggest that God's omniscience is not and cannot be true?

Let me stress first that God does not exist in time, as we do. The world has past, present, and future events, but God does not. So He does not have to wait until something happens in His creation in order to know that it happens. His being all-knowing means that

[137] Feser, *Five Proofs*, 212.

He knows everything by virtue of being the First Cause of everything. As Edward Feser said, "It is in a single, timeless act that God causes to exist everything that has been and will be."[138] Seen this way, the image of God watching a video is completely inadequate.

Besides, in your question, you keep creating a stalemate or deadlock between God's omniscience and human freedom. You are thinking in terms like these: If God infallibly knows that I will go on vacation next week, how could I possibly not go on vacation next week? Not going next week would mean that God had been wrong in thinking I would go—and God cannot be wrong. But if not going next week is impossible, then I cannot freely choose whether to go next week—my free choice has gone out the window.

There is something wrong with this scenario, though.[139] Suppose I know—through what you told me, through a hotel reservation, or from another source—that you will be on vacation next week. Obviously, my knowing this isn't incompatible with your decision to be on vacation next week. So why could not God know this too?

There is nothing illogical or even contradictory here. Let me explain this with Zeno's famous motion paradox about Achilles and the tortoise. As recounted by Aristotle, "In a race, the quickest runner can *never* overtake the slowest, since the pursuer must first reach the point whence the pursued started, so that the slower must always hold a lead."[140] Does this paradox prove that motion doesn't exist? No, it doesn't; instead it proves that motion is not so easy to understand. In a similar way, your objection to God's omniscience does not prove that it doesn't or cannot exist, but only that God's omniscience is hard to understand. Let's try to understand it a bit better.

---

[138] Ibid., 212.
[139] See more details in ibid., 213.
[140] *Physics*, bk. 6, pt. 9, 239b15.

A Catholic Scientist Proves God Exists

When I choose to have coffee, I supposedly follow my free will. But when the coffee machine makes coffee, it could not have acted differently—it just follows certain laws of nature that were applied in the machine by its makers. That's why the outcome of a coffee maker is predetermined. Although the machine is programmed to make coffee, I am not. What *both* cases have in common, however, is that God knows in advance what the outcome of either case is. In other words, God is the First Cause that makes both kinds of actions possible. Just as the machine could not make coffee without God, so could I not decide to have coffee without God. In Feser's words, "Just as with natural causes, if free choices were not caused by God, they couldn't exist at all."[141] Being foreseen by God doesn't make free actions less free.

This means that God's omniscience works in such a way that what God knows does not act upon what I do, as if God's omniscience were a secondary cause. God's knowing my decision is not the direct cause of my decision—I am—but He is the indirect cause that lets me be the cause of my own decisions. Seen in this light, our human freedom need not be in conflict with an all-knowing God. He is the complete and First Cause of a free act, whereas the human agent is its complete secondary cause. God's knowing that I decide as I do does not make my decision God's.

I still maintain that if God's omniscience entails that God can know "ahead of time" what I will decide, then my decision has already been made before I make that decision. You make it look as if God is reading the newspapers of the future, but without my playing any role in making the news.

As creatures, we are certainly familiar with *past* and *future*, but *past* and *future* do not apply to an eternal God. If there were

---

141 Edward Feser, *Aquinas* (London: Oneworld Publications, 2009), 151.

something God did not know yet, His knowledge would be imperfect. His knowledge does not change, analogous to the way mathematical truths do not change. Humans can change their minds, but God cannot change His mind, for that would involve a contradiction—namely, that His perfect knowledge would also have an imperfect part that reconsidered things He may have missed the first time around. If God were not all-knowing, He would not be God.

Therefore, God's position with respect to time is such that, unlike us, He does not have to wait for the future to unfold in order to know its contents. As we found out earlier, God is in no way a temporal being, but He is the Creator of time, with complete and equal access to all of its contents. If He were in time, even God could not know what has not yet happened—for to think differently would create a contradiction. But if God exists entirely outside of time—in a kind of eternal present to which all that occurs in time is equally accessible—He would indeed be able to comprehend all of history, the past and the present as well as the future, just as though they were now occurring. For this reason, an all-knowing God must have "foreknowledge" regarding future events, decisions, and actions because His foreknowledge is situated outside of time, external to time. It comprises all that ever was and ever will be.

God does not have to wait on the contingent and temporal event of a person's decision to know what that person's action will be—God knows it from eternity. St. Thomas argues that God's infallible knowledge of our future is not some secret predictive power.[142] Rather, in God's transcendent eternity, which is outside the flow of time, all events from any time are present to Him in one "eternal now"—regardless of whether they involve the past,

---

[142] *Summa Theologica*, I, Q. 14, art. 3.

present, or future. But seen from our "temporary now" perspective, we are still free actors who freely decide to cause things to happen.

Therefore, we must conclude that God is not only invisible, all-perfect, all-powerful, and all-present, but also all-knowing. At the same time, we must add to this that God is infinitely greater than all His attributes combined.

12

Why God Is All-Good

God's being *all-good* is another rational conclusion following from the proofs of God's existence. According to Aquinas, God as Pure Being Itself must also be good, indeed the highest good.[143] When we say "God is good," the meaning is not that God is the cause of goodness, or that God is not evil, but that, in the words of St. Thomas, "whatever good we attribute to creatures, pre-exists in God, albeit in a more excellent and higher way."[144] This is called God's benevolence.

Calling God all-good assumes that we know what goodness means. The word *good* is one of those ambiguous words our language is full of. It is usually associated with personal preferences, which makes it a rather subjective notion. Edward Feser uses a helpful example to show that this association puts us on the wrong foot.[145] A "good" Euclidean triangle must have three straight sides and three angles that add up to 180 degrees. That is not a matter of feelings, emotions, or preferences, but an objective requirement. There is nothing subjective about what makes a triangle good.

---

[143] *Summa Theologica*, I, Q. 6.
[144] *Summa Theologica*, I, Q. 13, art. 2.
[145] Feser, *Five Proofs*, 216–223. (I used several of his insights.)

A Catholic Scientist Proves God Exists

What does this have to do with our discussion of goodness? Most triangles drawn on paper have flaws, no matter how subtle, which make them bad triangles to some extent. The "badness" of something is a failure to be what the nature or essence of that something is. A triangle is good to the extent that its sides are straight and its angles add up to 180 degrees.

This means that the badness of something is a *failure* to be what its nature requires; it is the *absence* of what is required. This makes goodness different from badness in a fundamental way. Goodness is something present, whereas badness is something missing. The former refers to the presence of some feature, whereas the latter refers to the absence of that feature. The former is *actual*; the latter is not. But what they have in common is that they are both *real*. Blindness, for instance, is the absence of sight, but that doesn't mean blindness is not real. So the absence of goodness is not to deny the reality of badness.

How does God tie in with all of this? If *goodness* means to be actual in some way, and badness is the failure to be that way, then God, who is all-presence and pure actuality, must be pure goodness, too, with the absence of any sort of badness. In other words, God is all-good: there is nothing bad, or missing, in God.

That sounds nice, but it may actually be a dangerous conclusion, because it raises a very disturbing question: How can an all-good God be reconciled with the existence of evil in the world? Evil is the big elephant in your room. Evil is something very real, and yet bad.

You are right; now we are getting to the heart of the matter. Your question is fundamental, but certainly not new.[146] The skeptic philosopher David Hume worded the problem this way: "Is God

---

[146] The Greek philosopher Epicurus worded the question already around 300 B.C.

willing to prevent evil, but not able? Then he is impotent. Is he able, but not willing? Then he is malevolent. Is he both able and willing? Whence then evil?"[147] Evil has certainly been one of the strongest obstacles to Judeo-Christian faith. Does Hume have a point here, declaring God more malevolent than benevolent?

I would say that Hume is definitely wrong, and here is why. Let's start with what is called *moral evil*. Where does moral evil come from? I think we can be very brief in answering this question: moral evil stems from the fact that human beings are endowed with free will. They have decision-making power, given to them by the First Cause. By becoming free causes of their own, humans do not take anything away from God's sovereignty, but they exercise the free will that God has given them, even the freedom to do evil. Just as God makes His creatures movable by nature so that they can move themselves, so God gives them free will and decision-making power, thus allowing them to become causes of their own, which can lead to good or bad results.

The rhetorical question is: How could God ever give us freedom without accepting its consequences up to the point of our freely choosing any evil outcome? Why did the all-good God decide to do this and thus open the door for moral evil to come in? God could perhaps have chosen to eliminate the possibility of evil and evildoing, but then He would have also taken away the possibility of good and doing good—as well as the possibility of free choices. Free creatures are of greater value than puppet creatures, because their greater likeness to God is an improvement to creation—otherwise they would no longer be free agents, but only puppets manipulated by God. So we must conclude from this that moral evil doesn't come from God, who is all-good, but it originates in human beings. God did not introduce moral evil—*we* did.

---

[147] David Hume, *Dialogues concerning Natural Religion*, pt. 10.

I think you took the easiest case of the two first. We don't have to blame God for what human beings do wrong. But the case is much more daunting when it comes to *physical* evil — evil not caused by human beings, but rather evil that seems to be part of nature — such as natural death, famine, diseases, cancers, earthquakes, tsunamis, and other catastrophes.

Yes, that seems to be a much tougher case. Yet, several answers have been given as to where physical evil comes from if God is all-good.

A first answer would run as follows. The universe is bound to follow its God-given laws of nature, for the laws of nature must operate as they do. God could indeed have created a perfect world in which He was in control of everything so there would be no secondary causes, no contingency, and no imperfections. Instead, says Aquinas, it was in God's wisdom to ordain not to be responsible for all contingent events.[148] God allows defects in secondary causes — such as a falling stone hitting your head or an earthquake destroying your home — to exist. The evolutionary biologist Francisco Ayala rightly places this in a wider context: "As floods and drought were a necessary consequence of the fabric of the physical world, predators and parasites, dysfunctions and diseases were a consequence of the evolution of life."[149]

This explanation lets such events happen without making God directly responsible for any so-called imperfections or defects in the universe. As Michael Augros puts it, "If secondary causes, unlike the Primary Cause, are not infallible, if they are defectible, then we might well blame them, rather than the first cause, for any flaws we find (or think we find)."[150]

---

[148] *Summa Theologica*, I, Q. 22, art. 2.
[149] Francisco J. Ayala, *Darwin's Gift to Science and Religion* (Washington, D.C.: Joseph Henry Press, 2007), 5.
[150] Augros, *Who Designed the Designer?*, 103.

A second answer for the existence of physical evil is that, if there were no physical evil, such a fact would diminish the good of the universe. To explain this seemingly enigmatic answer, St. Thomas uses the example of the lion, which could not live without killing its prey. As he puts it, "If all evil were prevented, much good would be absent from the universe. A lion would cease to live, if there were no slaying of animals."[151] Whatever may be evil for the individual, the prey, is good for the larger picture, the universe. We might also add the example of pain: pain is certainly painful in itself, but it is also part of a larger good—a warning sign of something wrong.

Consequently, if there were no physical evil, such a fact would diminish the good of the universe as a whole. Environmentalists are very aware of this fact; even "dangerous" animals such as poisonous spiders and snakes play an essential role in their ecosystem; taking them out would disrupt the system. There is always a delicate balance between what looks "bad" in detail at first sight and what is "good" in the larger picture.

A third answer introduces an important distinction between what God *wills* and what God *allows*. Aquinas says that God "neither wills evils to be nor wills evils not to be; he wills to allow them to happen."[152] God does not will earthquakes, but he allows them when they are a consequence of the laws of nature—in the same way as God does not will wars but allows them when humans use their free will to start them. In other words, there is God's positive or *providential* will, and then there is His *permissive* will; therefore, not everything that happens in this universe is directly willed by God's benevolence. To say that God allows or permits evil does not mean that He sanctions it in the sense that He approves of it, or even wants it.

---

[151] *Summa Theologica*, I, Q. 22, art. 2. ad. 2.
[152] *Summa Theologica*, I, Q. 19, art. 9, ad. 3.

A fourth answer adds an entirely new dimension to the problem of both moral and physical evil. It can be found all over the Bible. It tells us there are demons in the world, which were created by God as angels with free will, just like us. But these angels decided against serving God. As John Milton's *Paradise Lost* puts it, Lucifer was banished from heaven when he said, "Better to reign in Hell than serve in Heaven."[153]

Satan played a decisive role in Paradise, and he keeps doing so as Jesus tells us in the parable of the weeds: "The Kingdom of heaven may be likened to a man who sowed good seed in his field. While everyone was asleep his enemy came and sowed weeds all through the wheat, and then went off.... His slaves said to him, 'Do you want us to go and pull them up?' He replied, 'No, if you pull up the weeds you might uproot the wheat along with them.'"[154]

No wonder Christianity sees the history of humanity as a perpetual, cosmic warfare between God and Lucifer, between the Light of God and the darkness of Satan, between an all-good God calling us to be His image and an all-evil Satan enticing us to be our own image, which is Satan's image.

A fifth answer tells us that God's creation is not finished yet but is in a state of journeying toward an ultimate perfection yet to be attained.

Let me interrupt you here. If God did not finish completely what He had created, then God is to be blamed for evil and for everything that's still missing in this world. Isn't that the only sensible conclusion?

Not at all. God may have had a good reason for not completely "finishing" what He had created. Some goods would be eliminated

---

[153] John Milton, *Paradise Lost*, bk. 1, lines 242–270.
[154] Matt.13:24–30.

if God had decided to get rid of all deficiencies from the very beginning. The good of having a free will, for example, would have been eliminated if God had eliminated the possibility of humans freely deciding on actions that are evil. Aquinas puts this in the right perspective—an all-good being could possibly have a reason to allow evil and to produce good out of it.[155]

God's creation is not perfect yet, but it is on its way. And human beings have been made participants in this process; they are God's "coworkers" in bringing His creation to perfection. So physical evil is part of the imperfection that still surrounds us.

This last explanation is not some outlandish answer to the problem of evil in creation. The *Catechism* puts it very emphatically:

> Creation has its own goodness and proper perfection, but it did not spring forth complete from the hands of the Creator. The universe was created "in a state of journeying" (*in statu viae*) toward an ultimate perfection yet to be attained, to which God has destined it. We call "divine providence" the dispositions by which God guides his creation toward this perfection. (no. 302)

The bottom line of this quote is that human beings have been entrusted "with the responsibility of 'subduing' the earth and having dominion over it (cf. Gen. 1:26–28).... They then fully become 'God's fellow workers' and co-workers for his kingdom (1 Cor. 3:9; 1 Thess. 3:2; Col. 4:11)" (CCC 307). The key point is that God has given us a role in His creation.

Nonetheless, most of us have this rather common perception that a perfect world from the outset would have been better than an imperfect world on its journey to perfection.

---

[155] *Summa Theologica*, I, Q. 2, art. 3.

I would reply that this common perception is actually a misperception. We tend to assume that God's providence should have created an ideal world, a hedonistic paradise, a place in which comfort and convenience are maximized, a world in which everyone has an electrode implanted, so to speak, to cause intense euphoria and ecstasy with a simple push of the button.

But that assumption stands in contrast to some strong feelings and intuitions most of us have. Do we really admire those who appear to have a life of ease? What we admire instead are lives of courage and sacrifice; we have a high regard for people who overcome hardship, deprivation, or weakness so as to achieve some notable success. We admire people who stand against some great evil or who relinquish their happiness to alleviate the suffering of others.

As a matter of fact, a life without challenge is a life without interest. Since our youth, we have been conditioned to view suffering as an impediment to happiness. We live in a world that runs away from suffering; our bathroom cabinets are filled with painkillers. Yet we could be missing out on another dimension of suffering, for suffering has the mysterious potential of redeeming us, transforming us, transfiguring us. Suffering can be very therapeutic. Whereas the Stoics say, "Suffering is nothing," Christians say, "Suffering is everything." Apparently, the maximization of creaturely pleasure does not seem to be a top priority in most lives — let alone in God's omnipotence and benevolence. An all-good and all-powerful God has the power even to turn something bad into something good.

Remarkably, the first chapter of the book of Genesis calls God's creation *good* several times, but it does not use the word *perfect*. God's creation is not perfect, but it is on its way to perfection — and we human beings have been assigned an active role in bringing God's creation to perfection. God *wills* perfection but *allows* imperfection on the journey to perfection. So physical evil is part of the imperfection by which we are still surrounded. But the all-good God has only good in mind for us, no matter what we see right now. The book of Genesis tells us that when Joseph, who had been sold into slavery by his brothers and taken to Egypt, was finally able to see his brothers again, he said to them, "Even though you meant harm to me, God meant it for good."[156] God draws good from evil.

So we cannot say that evil was put in the world by an all-good God. Thomas Aquinas used to say, "You cannot give what you do not have." God is all-good, so He cannot give what he doesn't have—evil. The physical defects and failures we see in this world did not directly come from an all-good God. They are perhaps something like side effects or imperfections, but not directly "willed" by God.

You keep talking about what God wills, but I keep wondering where God's will comes from. It doesn't seem to be one of His attributes.

One might say that God's will comes from His being all-good. But it could also be said that it comes from His being *all-loving*. God is Love Itself, whose goodness desires to give itself away and to invite other beings to share in His being. God loves what He created—love is the "passion" that made God create. He loves every part of what He created—that's why He created it.

We tend to think of love as a mere feeling or a sheer emotion. But love is a matter of one's will, which is something active, rather than of one's emotions, which are passive. One who loves actively wills what is good for the one being loved, even if the feeling may not always be there. Love can't be self-contained, or it isn't really love. Love is something that goes out and becomes an interaction. This also holds for God's love. God loves all that He creates, and He wills what is good for all that He creates. That's the reason we

---

[156] Gen. 50:20.

must also attribute love to God.[157] Being all-good and being all-loving are intimately united in Him.

Put differently, God created the universe not out of personal neediness but out of love and generosity. Thus, God also conferred love and goodness on human beings, made in His image. This means that love is an essential part of us, and we should share love and goodness with others as well. Perhaps that makes us realize that love and goodness point to a far superior love and goodness coming from the Creator, who created all there is.

---

[157] *Summa Theologica*, I, Q. 20.

13

Why God Is Triune

All the attributes of God we have discussed so far—being all-perfect, all-powerful, all-present, all-knowing, all-good, and all-loving—can be derived by reason alone and can be proven in a deductive way, regardless of any specific religion.

Other things about God, however, cannot know by using reason alone. One of them is that God is *trinitarian* or *triune*—Father, Son, and Holy Spirit—which is *not* a conclusion based on mere reason. We are dealing here with the God of faith, not the God of reason. This does not mean, of course, that we can say anything we want about the God of faith. The reason for this caveat is simple: even religious faith is something about which we can reason, argue, and debate. It was said about St. Paul that "he entered the synagogue, and for three months debated boldly with *persuasive* arguments about the kingdom of God."[158] That means we should still be able to reason, argue, and debate about the triune God of faith.

There is always a delicate balance between faith and reason. St. Augustine could not have said it more clearly: "Believers are also thinkers: in believing, they think and in thinking, they believe...

---

[158] Acts 19:8 (emphasis added).

If faith does not think, it is nothing."[159] He also introduced his two famous formulas, which express the coherent synthesis between faith and reason: "Believe that you may understand" (*crede ut intelligas*), but also, and inseparably, "Understand so you may believe" (*intellige ut credas*).[160]

The same idea can also be found in St. Anselm's two famous phrases. His first phrase says "I believe so I may understand" (*credo ut intellegam*),[161] which underscores the role of faith for better understanding. Faith is often necessary for us to understand the world and ourselves better. C. S. Lewis summarized this beautifully when he wrote, "I believe in Christianity as I believe that the sun has risen: not only because I see it, but because by it I see everything else."[162]

Anselm's second well-known phrase is of equal importance. It speaks of "faith seeking understanding" (*fides quaerens intellectum*),[163] and this stresses the role of reason to explain faith. Faith needs reason for us to understand faith better. It also tells us that faith cannot be "unreasonable." If something in our faith goes against reason, God cannot endorse it, let alone create it. Both phrases taken together stress the unity of faith and reason.

How do we take this twosome? On the one hand, knowing by faith is different from knowing by reason. What we know by faith about God came to us through God's special revelation—that is, through the Bible and Christian Tradition. When St. Paul was in Athens and saw an altar inscribed "To an Unknown God," he invited the Athenians to take the step from a "God of reason" to a "God of faith." He told them, "What therefore you worship as

[159] Augustine, *De Praedestinatione Sanctorum*, 2, 5.
[160] Augustine, *Sermon* 43, 7, 9.
[161] Anselm, *Proslogion* I.
[162] C. S. Lewis, "They Asked for a Paper," in *Is Theology Poetry?* (London: Geoffrey Bles, 1962), 165.
[163] Anselm, *Proslogion* II.

unknown, this I proclaim to you."[164] It is only by faith and God's special revelation that we know God is triune. Although a truth like this about God cannot be known by reason alone, reason must come back in to show that our faith is not *against* reason.

On the other hand, what knowing by faith and knowing by reason have in common is a factual basis. What we know about God through reason and what we know about God through faith is not just a matter of personal beliefs or subjective opinions—instead, both kinds of knowledge are about *facts*. Why should we believe something? Because others believe it? No! Because it feels good? No! The only reason is because it is *true*.

How do we know it's true? The fact that a religion claims to be true doesn't make that religion true. Similarly, when defendants plead innocent, that doesn't mean they are innocent.

You are right; there must be some way to validate a religion. I would say there are at least two.

One way is to do what we did in the previous chapters: move from true premises to true conclusions by using mere reason. God either exists or He doesn't—that's not a matter of opinion, but a mere fact. Another way is to listen to what God reveals to us about Himself through special revelation. God is triune, or He is not—that is a fact, not a matter of opinion (although it is a fact we could not have known without revelation). If God is not triune, then our facts are wrong, as is our faith. Faith cannot change the facts, just as a belief in a flat earth cannot make the earth flat.

This makes *reason* our most important tool to decide what is true and what is false. Religious beliefs that are against reason cannot be true; religious faith that is not open to reason cannot be true; religions that are irrational and incoherent cannot

---

[164] Acts 17:23.

be true; contradictions between religions cannot be accepted. In fact, reason is the best, if not only, criterion to judge a religion's truth. Reason acts like a litmus test for religions. Religions based on extraterrestrial sources or on books that only some people are supposed to have access to or on books that no longer exist have a hard time passing the test of reason.

As a consequence of this, we should never reduce "religion" to a mere set of beliefs and opinions, *untested by reason*. Once we abandon reason as a litmus test for religion, anything can go under the banner of "beliefs"—even white-supremacist beliefs would qualify as "religion." In other words, not everything that calls itself religion should be regarded as a legitimate form of religion.

This litmus test should also be applied to the triune God of the Christian religion. Judaism and Islam, for instance, teach their followers that Jesus is not the Son of God, whereas Christianity teaches that He is. Well, Jesus either is the Son of a triune God or He's not—these two claims can't both be right. If Jesus is indeed the Son of a Triune God, that is a fact we cannot deny. The mere belief that He is *not* the Son of a triune God cannot change the facts.

If Christians claim that God is indeed a Triune God, then they need to explain why that claim is not against reason. What does it mean that God is triune? Doesn't the Jewish Shema Yisrael[165] say "Hear, O Israel: The Lord is our God, the Lord is *one*"? Doesn't this imply God is not *three*?

It's not quite clear how the last part of your quote should be translated—either as "the Lord is *one*" or as "the Lord *alone*."[166]

---

[165] Deut. 6:4.

[166] The Hebrew word *echad* means "one," but sometimes it means "alone" or "only" (see Josh. 22:20; 1 Chron. 29:1; Isa. 51:2). Hebrew doesn't have a form of the verb *to be* to link a subject with a

The former version puts the emphasis on the oneness of God, and the latter on the sole worship of God to the exclusion of other gods. Both translations are theoretically possible, but the latter seems to be more in line with the permanent Jewish battle against the idols and false deities they were exposed to in their surroundings.

Yet the most common objection against a triune God is that it violates monotheism. But does it? People who think we are dealing with a logical or mathematical inconsistency might consider the following analogy: the Trinity does not represent $1+1+1=3$, which would amount to polytheism, but rather $1 \times 1 \times 1 = 1^3 = 1$, which is in accord with monotheism. Or, with the help of a geometric analogy: the fact that we distinguish three dimensions does not mean that we can separate those three dimensions. In a similar, analogous way, the Trinity is like the three "dimensions" of the same one reality, God. St. Patrick famously explained the doctrine of the Holy Trinity to his flock in Ireland by using the three leaves of the shamrock: each leaf represents one of the three persons, yet it is still only one shamrock. Or think of a triangle: it has three lines forming three angles, yet it is only one figure.

St. Augustine used another analogy to help us understand why I cannot be in love with love unless I love a lover; for there is no love where nothing is loved. So there are three things: the lover [Father], the loved [Son], and the love [Holy Spirit]."[167] Augustine's analogy is particularly striking when we realize that God's attribute of being all-loving even extends to the very inner being of the Holy Trinity. God is a triad of love: a going out in love, a

---

predicate. So "the Lord alone" and "the Lord [is] one" are technically both possible, though very different in meaning.

[167] "On the Trinity," in *Basic Writings of St. Augustine*, vol. 2, ed. Whitney J. Oates (Grand Rapids, MI: Baker Books, 1992), 790.

return in love, and thus, even more, love itself. God is all-loving, a perfect love—which requires three![168]

Each of the three Persons in the Trinity refers constantly to the other. The Trinity is an internal bond of love between Father, Son, and Spirit. This internal bond is not only an internalized feature of the Trinity, but it was also externalized when God's creation came into being. Thus, the paradigm of God's love in the Trinity became also the epitome for the love in our own lives, since we were made in God's image and likeness. We will never be all-loving like God, but we are called to be as loving as humanly possible.

In reference to the internal unity of the Holy Trinity, the *Catechism* explicitly states: "Christians are baptized in the *name* of the Father and of the Son and of the Holy Spirit: not in their *names*, for there is only one God, the almighty Father, his only Son and the Holy Spirit: the Most Holy Trinity" (no. 233). Put differently, in the Trinity, the Father is not the Son; the Son is not the Father; and the Spirit is neither Father nor Son. They are all "one" precisely by not being each other—one in being, yet diverse in person.

St. Hilary of Poitiers, one of the early Church Fathers, puts it quite paradoxically: "Each Divine Person is in the Unity yet no Person is the One God."[169] In short, there is *one* God but in three Persons. Because the three Persons are all "one" in being, they share the same divine attributes: each Person in the Trinity is all-perfect, all-powerful, all-present, all-knowing, and all-good.

If you are right that belief in a triune God is not against reason, then the question arises: How do we know that God is triune? To

---

[168] For a good explanation, see Tom Massoth, "Perfect Love Requires Three," Simple Catholic Truth, February 12, 2017, simplecatholic-truth.com/2017/02/12/perfect-love-requires-three/.

[169] Hilary of Poitiers, *On the Trinity* 7, 2.

say that Jesus told us so is not a satisfying answer. For then the question is: How did Jesus know?

You are right: the answer that Jesus told us so is not very convincing in and of itself. That answer can be true only if Jesus is indeed God. But how do we know Jesus is God? We can't just say Jesus is God because Jesus says so. Most people who say they are God end up in an asylum, as do people who claim to be Jesus. Ironically, the main reason for Jesus' crucifixion was his blasphemous claim of being God. It was a claim with a death sentence.

Yet that claim was fundamental for Jesus' mission. Only a person who comes and speaks from the "inside" can show and tell us what God is like. In St. John's Gospel, Jesus says, "Believe me that I am in the Father and the Father is in me."[170] Denying that Jesus is God has very serious consequences. If Jesus is not God but merely a human being, then His words and actions are worth as much, or rather as little, as anyone else's.

Another implication of denying that Jesus is God is perhaps even more detrimental: if Jesus was only *like* God, then someone else even *more* like God might arise somewhere. If that might happen, then Jesus was only an interim rather than a definitive revelation of God. If so, then the revealed truth so far might be less than that of later arrivals. If so, then all of Jesus' statements and commandments would have only temporary value. This would be contrary to what the apostles and the Church have been affirming over and over again—that Jesus was the definitive Word of God, that He was fully God and fully man, and that His teaching, life, death, and Resurrection were definitive, not interim, revelations from God.

C. S. Lewis, once an agnostic, understood this very clearly when he wrote about Jesus:

---

[170] John 14:11.

I am trying here to prevent anyone saying the really foolish thing that people often say about Him: "I'm ready to accept Jesus as a great moral teacher, but I don't accept His claim to be God." That is the one thing we must not say. A man who was merely a man and said the sort of things Jesus said would not be a great moral teacher. He would either be a lunatic—on a level with the man who says he is a poached egg—or else he would be the Devil of Hell. You must make your choice. Either this man was, and is, the son of God: or else a madman or something worse.... Now it seems to me obvious that He was neither a lunatic nor a fiend: and consequently, however strange or terrifying or unlikely it may seem, I have to accept the view that He was and is God.[171]

Jesus' being God may be fundamental to Christianity, but I wonder whether Jesus Himself ever said He was God. It has often been claimed that this idea came up afterward, and only within certain Christian communities.

Such a claim is hard to substantiate, though. Jesus was executed because He expressed the blasphemous claim of being God. When Jesus asked His attackers why they wanted to stone Him, they replied, "For a good work we do not stone you, but for blasphemy; and because you, being a man, make yourself out to be God."[172]

The Christian belief that Jesus was not only the Messiah but also the Son of God can be found all over the New Testament. Critics of this view like to stress that the most explicit expressions of this claim can be found only in the latest the four Gospels, St. John's. True, St. John is very categorical and persistent about Jesus' claim

---

[171] Lewis, *Mere Christianity*, bk. 2, chap. 4.
[172] John 10:33.

to be God. He says, for example, that Jesus existed before Abraham (8:58), and that Jesus was equal with the Father (5:17, 18).

But even in the oldest Gospel, St. Mark's, we find this belief of Jesus' divinity repeatedly proclaimed—for instance, at His baptism by John the Baptist, "And a voice came from the heavens, 'You are my beloved Son; with you I am well pleased,' "[173] and again during the Transfiguration: "From the cloud came a voice, 'This is my beloved Son. Listen to him.' "[174] Mark also mentions that Jesus claimed the ability to forgive sins (see Mark 2:5–7), which is something God alone can do, according to the Bible (see Isa. 43:25).

We find this claim of divinity, in fact, as early as the oldest writings of the New Testament, written before the Gospels: that is, in the letters of St. Paul, written in the 50s. In his oldest letter—the oldest document in the New Testament—St. Paul mentions already God's "Son from heaven, whom he raised from [the] dead, Jesus."[175] He also says about Jesus that in Him, God "was manifested in the flesh, vindicated in the Spirit, seen by angels, preached among the nations, believed on in the world, taken up in glory."[176] Paul also wrote, "But when the fullness of the time came, God sent forth his Son, born of a woman, born under the law."[177] And again, "Yet for us there is but one God, the Father, from whom all things are and for whom we exist, and one Lord, Jesus Christ, through whom all things are and through whom we exist."[178]

Unmistakably, the consistent testimony of Jesus and the writers of the New Testament is that Jesus was more than a mere man—He

---

[173] Mark 1:11.
[174] Mark 9:7.
[175] 1 Thess. 1:10.
[176] 1 Tim. 3:16.
[177] Gal. 4:4.
[178] 1 Cor. 8:6.

was God. So the claim that Jesus' divinity is a later development has no basis in Scripture but only in the minds of some critics.

The consequence of this is that God is more than one. God is at least two — Father and Son — but is actually three — Father, Son, and Holy Spirit. He is triune. Trinitarian expressions can be found all over the New Testament.[179] St. Matthew even mentions that Christians should be baptized in a Trinitarian way: "Go therefore and make disciples of all nations, baptizing them in the name of the Father, and of the Son, and of the Holy Spirit."[180]

To articulate the dogma of the Trinity, the Church had to use reason even more to develop her terminology with the help of certain notions of philosophical origin — notions such as *substance*, *person*, *hypostasis*, *relation*, and so on. With the help of these concepts, the Church could make clear that she does not confess three Gods, but one God in three Persons, the "consubstantial Trinity." Each of the three Persons in the Godhead possesses the same eternal and infinite divine nature; thus, they are not three Gods, but the *one* true God in essence or nature. The classic formula, articulated by the theologian Tertullian around 200, has been "one substance, three persons."

Obviously, we are dealing here with a mystery. God's being triune is one of the mysteries of faith. A *mystery* is not something about which we can't know anything, but something about which we can't know everything. Just because it is a mystery does not mean we shouldn't spend time thinking about the Trinity. Rather, it means that we can spend our entire lives thinking about it and never come to the end of it.

The reason there are mysteries is that God is infinite, which we know by reason, whereas our intellects are finite, which we know

---

[179] Including Acts 2:33; 1 Cor. 12:4–5; 2 Cor. 13:14; Eph. 1:13–14; 4:4–6; 1 Pet. 1:2; and Rev. 1:4–5.
[180] Matt. 28:19.

by experience. Although mysteries of faith may be beyond reason, they are not unreasonable; they can be explained and defended, although not proven, by arguments based on reason. Sometimes that requires heavy, precise terminology—often hard to digest.

The terminology you are referring to is probably what most Christians know as *dogmas*. That's what we find in the Nicene Creed, for instance. But my problem is that dogmas have a bad name: they are dull and hard to understand.

The Nicene Creed may appear to us like a case of theological hair-splitting, yet it was an issue Christians were willing to die for. That explains why dogmas can create so much rioting, fighting, exiling, and even killing. The truth, however, may very well be in the details. The Creed is the "Faith of our Fathers" for which Christian citizens of the Roman Empire suffered imprisonment, torture, and death. G. K. Chesterton once said, "Truths turn into dogmas the instant that they are disputed."[181] And the late, great English writer Dorothy Sayers would add to this that the drama is in the dogma, for which our ancestors were willing to die:

> Official Christianity, of late years, has been having what is known as bad press. We are constantly assured that the churches are empty because preachers insist too much upon doctrine—dull dogma as people call it. The fact is the precise opposite. It is the neglect of dogma that makes for dullness. The Christian faith is the most exciting drama that ever staggered the imagination of man—and the dogma is the drama.[182]

---

[181] *Heretics* (New York: John Lane Company, 1905), 304.
[182] Dorothy Sayers, "The Greatest Drama Ever Staged," in *The Whimsical Christian* (New York: Macmillan, 1978), 11.

14

Why God Is Incarnate

The late Supreme Court judge Antonin Scalia once told the story
of the best lesson he ever learned during his studies at Georgetown University.[183] It happened during his oral comprehensive
examination at the end of his senior year. His history professor,
Dr. Wilkinson, asked him one last question: "Of all the historical
events you have studied, which one in your opinion had the most
impact upon the world?" Scalia mulled over several options: the
French Revolution, the Battle of Lepanto, or perhaps the American Revolution? Whatever he eventually happened to answer, Dr.
Wilkinson informed him of the correct answer. Of course, it was
the Incarnation—the moment God came to Earth in the person
of Jesus Christ. That's what he considered the one real turning
point in the history of Planet Earth!

The doctrine of the Incarnation—an infinite God becoming a
finite man in Jesus of Nazareth—is probably even more of a mystery
than a triune God. It seems to go against all the divine attributes
of God, perhaps even against reason. Fortunately, although reason
can never prove the Incarnation all by itself, reason can still help

---

[183] Antonin Scalia, *Scalia Speaks: Reflections on Law, Faith, and Life
Well Lived* (New York: Crown Forum, 2017), 147.

us understand what the term means. Obviously, there could not be an Incarnate God with a triune God.

I think some rationality and reasoning is needed here. A God becoming man looks very similar to what pagans say about their deities.

Yes, we certainly need some reason to explain what an incarnate God is like. First of all, it is not the triune God who became man in the Incarnation. It is just one of the three Persons of the Trinity, the Son of God, who became man—God incarnate. As said earlier, Jesus is God, but God is not Jesus. It's not the triune God or the Trinity that became incarnate—that would indeed be against reason. The Trinity is the full divine being, which is all-perfect, all-powerful, all-present, all-knowing, all-good, and all-loving. The Trinity is the unity of three Persons who are all "one" precisely by not being each other, so they all share the attributes of the triune God. The same can be said about the incarnate God.

Second, the Incarnation speaks of a delicate balance between Jesus' divinity and His humanity—Jesus was truly God and truly man, fully divine and fully human. Jesus has two essential "poles" of being, so to speak. The *Catechism* says that the Incarnation of the Son of God "does not mean that Jesus Christ is part God and part man, nor does it imply that he is the result of a confused mixture of the divine and the human. He became truly man while remaining truly God. Jesus Christ is true God and true man" (no. 464).

How do we balance these two sides? They are equally essential. On the one hand, had Jesus not been *human*, then the Crucifixion was not real but only an illusion. As Pope St. Leo the Great put it in 451, "Invisible in his own nature, he became visible in ours. And he whom nothing could contain was content to be contained."[184]

---

[184] Sermon 22 on the feast of the Nativity II, 2, in *Nicene and Post-Nicene Fathers*, 2nd series, vol. 12, *Leo the Great, Gregory the Great,*

So Jesus' birth, Crucifixion, and death were real, not fake. They came with Jesus' being human.

On the other hand, had Jesus not been *divine*, then He would have been only a prophet at best, and the whole economy of salvation would be up for grabs. No mere man can take away all the sins of the world[185] — only a God-man can. St. Proclus of Constantinople said in 429, "We do not proclaim a deified Man, but we confess an incarnate God."[186] Jesus could be our Savior only by being the Son of God, an incarnate God, both divine and human. No one else could take away the sins of the world.

The Incarnation is tightly connected with God's being all-loving. St. John expresses this emphatically: "For God so loved the world that he gave his only Son, so that everyone who believes in him might not perish but might have eternal life."[187] *Love* is the key word here; it's the love of an all-good, all-loving God. God so loved what He made — all of creation, all of us — that He came down in His Son, Jesus, to redeem all of us, which is something only a incarnate God can do.

Redemption is for all of us, because Christ "died for all," in the words of St. Paul.[188] The *Catechism* puts it very emphatically: "There is not, never has been, and never will be a single human being for whom Christ did not suffer" (no. 605). Jesus' coming down from Heaven flows from God's being all-loving. In His divinity, Jesus was able to redeem all of us by suffering in His humanity.

No wonder St. Paul speaks about Christ Jesus in very dramatic terms that cannot be misunderstood:

ed. Philip Schaff and Henry Wace (New York: Christian Literature, 1895), 130.

[185] In reference to John 1:29.
[186] Proclus of Constantinople, *Sermon on the Annunciation*.
[187] John 3:16.
[188] 2 Cor. 5:15.

Christ Jesus who, though he was in the form of God, did not regard equality with God something to be grasped. Rather, he emptied himself, taking the form of a slave, coming in human likeness; and found human in appearance, he humbled himself, becoming obedient to death, even death on a cross.[189]

Let's stress it again, there are two sides of Jesus: a divine side and a human side. Many have given Him other sides. Most notably in the past century, some theologians began to create rather different portrayals of Jesus. They painted Him in various exclusive but mutually contradictory ways—Jesus as the anti-Roman revolutionary, Jesus as liberator of the poor, Jesus as the meek moral teacher, Jesus as a visionary, Jesus as an immigrant, Jesus as a radical, and so on. But as Pope Benedict XVI said in one of his sermons about all these portrayals, "they are much more like photographs of their authors and the ideals they hold."[190] Nowadays we would say they are more like "selfies" of these authors. Their makers eliminate anything about Jesus that does not fit their theology and image of Jesus. The Bible, on the other hand, describes Jesus from His two primary sides: His divine nature and His human nature. The rest flows from that.

The delicate balance between these two sides is also vital to understanding the significance of Jesus' suffering on the Cross. It was Jesus' divine nature that gave His Crucifixion a universal significance. But let's not misunderstand this: it doesn't mean God Himself was nailed to the Cross on Golgotha. God is more than Jesus—Father, Son, and Holy Spirit. During Jesus' suffering, the Son of God was in pain, but God the Father was in charge. To think that whatever happened to the Son happened to the Father

---

[189] Phil. 2:5–8.
[190] *Jesus of Nazareth*, trans. Adrian J. Walker (San Francisco: Ignatius Press, 2008), xii.

is actually considered a heresy, because it is based on the erroneous idea that there is only one Person in God's nature. It is in essence an anti-Trinitarian heresy. It was not the triune God who was nailed to the Cross but rather the Son of God who had become man.

Therefore, there are two sides also to what happened on Calvary. On the one hand, to hold that the Son suffered only in His *divine* nature would mean that Jesus did not experience suffering the way human beings do. In other words, to place the significance of the Son's suffering within His divine nature is to relegate His human suffering to insignificance and thus to demote all human suffering to insignificance.

On the other hand, to hold that the Son suffered only in His *human* nature would mean that He suffered only like every other human being suffers, which would make His suffering inadequate for the redemption of all of humanity. The suffering of the man Jesus would not be worth more than anyone else's suffering — only an incarnate God, a God-man, can redeem all of us.

To keep this delicate balance in place, it needs to be stated that God the Father "suffered with" His Son in the Crucifixion — not by becoming a God in pain on the Cross of His Son, but by remaining a God in charge who watched the suffering of His Son with divine love and then raised Him from the dead. As Pope Leo the Great said about Jesus' suffering, "Being God who cannot suffer, he did not disdain to be man that can suffer."[191]

It's certainly a very delicate balance. Not surprisingly, one group of heretics in the history of Christianity went to one extreme: Jesus is God but not really man, divine but not human. They could not accept that the Incarnation made Jesus Christ one of us — a real man, including all the limitations of time and space associated with His earthly humanity. This became known as the heresy of

---

[191] Sermon 22, in *Nicene and Post-Nicene Fathers*, 12:130.

Docetism, in which Jesus Christ only "appeared" or "seemed" to be a man, to have been born, to have lived and suffered and died.

St. Paul was reacting to this heresy when he wrote, "For in him dwells the whole fullness of the deity *bodily*."[192] And St. John refers to the same heresy: "Many false prophets have gone out into the world. This is how you can know the Spirit of God: every spirit that acknowledges Jesus Christ come in the *flesh* belongs to God."[193]

Another group of heretics in the history of the Church went to the other extreme: Jesus is man but not God, human but not divine. This is basically the position of a heresy that became known as Arianism. Whereas the previous heresy denied Jesus' true humanity, the heresy of Arianism denied His true divinity.

Arians came to this point by making some seemingly "logical" statements: if God is one, then Jesus cannot be God as well; if Jesus is the Son of God and also God Himself, then He cannot also be His own son (besides, what son is as old as his father?); if Jesus is Son, then He was begotten; if He is begotten, then He had a beginning; if He had a beginning, then He is not infinite; if He is not infinite, then He is not God; if there was a time when the Son was not the Son, then He had to be created at some point. In logic and math, that would be followed by *Quod erat demonstrandum* (QED). It's as simple as that! Or is it?

I am sure the Church, the staunch defender of reason, wanted to deal quickly with these Arians and with their logical questions. Answers were much needed. What could they come up with?

They came up with the best they could get through the use of reason in order to keep the core message of the good news protected against misunderstandings.

---

[192] Col. 2:9 (emphasis added).
[193] 1 John 4:1–2 (emphasis added).

Where did the Arians go wrong? For one, they considered the terms *begotten* and *created* synonymous, which is a very questionable equation. C. S. Lewis explains the difference between *begotten* and *created* lucidly, as only he can:

> When you beget, you beget something of the same kind as yourself.... But when you make, you make something of a different kind than yourself.... What God begets is God; just as what man begets is man. What God creates is not God; just as what man makes is not man.[194]

But a lot more had already been said by the Church long before Lewis entered the scene. St. Athanasius of Alexandria is arguably one of the earliest and strongest foes of Arianism. After Athanasius left the Egyptian desert, where he had lived among the hermits, he brought with him a book he had completed there around 318, with the title *On the Incarnation*. C. S. Lewis would later praise it for the "classical simplicity" of its style and called it a "masterpiece."[195] The book is addressed to a recent convert named Macarius, who had heard some confusing objections made against the deity of Christ, probably of Arian origin. It contains many jewels of orthodox Christian doctrine.

In this book, Athanasius made it crystal clear that Jesus is truly God and truly man, both Son of God and Son of Man. He came down from Heaven, God from God, light from light, taking on a human form, made of flesh and blood, to become our Savior. In Athanasius's words:

> He took pity on our race, and had mercy on our infirmity, and condescended to our corruption, and, unable to bear that

---

[194] C. S. Lewis, *Mere Christianity*, bk. 4, chap. 1.

[195] C. S. Lewis, introduction to Saint Athanasius, *On the Incarnation* (Yonkers, NY: St. Vladimir's Seminary Press, 2012).

death should have the mastery—lest the creature should perish, and His Father's handiwork in men be spent for nought—He takes unto Himself a body, and that of no different sort from ours.[196]

Athanasius saw in all clarity that the divinity of Christ is the cornerstone of our salvation. If Christ were only a creature, the gospel would not truly be such good news after all. Jesus' suffering would end up being worth as much, or as little, as anyone else's suffering. Creation can be renewed and redeemed only by its Creator—that is, by the divinity of Jesus. No human beings could accomplish this on their own. Only God can forgive sins.

Athanasius worked very hard to combat the Arian fallacy that the flesh of the Son of Man deprived Him of His divinity as the Son of God:

> For the Flesh did not diminish the glory of the Word; far be the thought: on the contrary, it was glorified by Him. Nor, because the Son that was in the form of God took upon Him the form of a servant was He deprived of His Godhead. On the contrary, He has thus become the Deliverer of all flesh and of all creation. And if God sent His Son brought forth from a woman, the fact causes us no shame but contrariwise glory and great grace. For He has become Man, that He might deify us in Himself.[197]

Athanasius must have had a hard battle to fight the seemingly logical points of the Arians. Did his efforts pay off?

[196] *On the Incarnation*, 8, 2. For translations of the writings of the Church Fathers, we used the "thirty-eight volume set," published between 1867 and 1900. These texts can also be found on the New Advent website: http://www.newadvent.org/fathers/.

[197] *On the Incarnation*, 60, 4.

The crowning glory of his work was the moment when Emperor Constantine had been convinced to summon an assembly of bishops to resolve the dispute with Arianism in the city of Nicaea in 325. Constantine had invited all eighteen hundred bishops of the Christian Church within the Roman Empire (about one thousand in the East and eight hundred in the West), but a smaller, unknown number attended. Eusebius of Caesarea counted more than 250, Athanasius of Alexandria counted 318, and Eustathius of Antioch estimated some 270 bishops.

Several bishops at the Council of Nicaea had brought along personal assistants. One of them was Athanasius, a deacon at the time, who accompanied Alexander of Alexandria, whom he would succeed later as patriarch of Alexandria. Dorothy Sayers, in her dramatized, but historically correct, 1951 play *The Emperor Constantine*, has Athanasius address the bishops at Nicaea in response to Arius, the founder of Arianism, as follows:

> Beloved Fathers, in whom will you believe? In the Christ of Arius, who is neither true man to bear our sorrows nor true God to forgive us our sins? Or in him who, being in the form of God, clung not to his equality with God, but was made in the likeness of man and became obedient unto death for our sakes?[198]

For good reason, Athanasius became known as "Athanasius against the world" (*Athanasius contra mundum*). In the Eastern Orthodox Church, he is called the "Father of Orthodoxy." Some Protestant groups call him "Father of the Canon." Without him, all Christians could have ended up Arians.

---

[198] Dorothy Sayers, *The Emperor Constantine* (Grand Rapids, MI: Eerdmans Publishing, 1976), 148.

A Catholic Scientist Proves God Exists

There is still a key point missing. With the world being God's creation, God could do with it as He pleased. If He wanted to forgive humankind, why couldn't God simply forgive what humanity had done wrong and thus redeem us?

That's a question we need to answer. Many theologians have tried to explain why God had to become man and die on the Cross for our salvation. Probably the most explicit "solution" was given by St. Anselm in his famous 1098 best seller *Cur Deus Homo* (Why did God become man?). Anselm basically argued that the insult given to God through sin is so great that only a perfect sacrifice could be acceptable to God, and that only Jesus, being both God and man, could be this perfect sacrifice. Anselm thought he could deductively prove that God had to come down to Earth in His Son, Jesus, to atone for our sins. To do so, however, he had to use several premises that cannot be accepted through reason alone. So his deduction could prove his conclusion only on the condition that one accepts the premises, which requires faith in itself.

Anselm was aware of how limited his proofs of the Incarnation were when he said about people who asked him what his proofs were worth: "They say that these proofs gratify them, and are considered sufficient. This they ask, not for the sake of attaining to faith by means of reason, but that they may be gladdened by understanding and meditating on those things which they believe."[199]

So the question keeps pressing for an answer. In a 2016 interview, Pope Emeritus Benedict XVI asked the same question: "Why the cross and the atonement?"[200] After mentioning all the horrific evils we have witnessed now and in the past, he continued,

---

[199] *Cur Deus Homo*, bk. 1, chap. 1.
[200] Benedict XVI, "The Beginning of Faith" *L'Osservatore Romano*, March 16, 2016.

This mass of evil cannot simply be declared nonexistent, not even by God. It must be cleansed, reworked and overcome.... God simply cannot leave "as is" the mass of evil that comes from the freedom that He Himself has granted. Only He, coming to share in the world's suffering, can redeem the world.... [Therefore,] when the Son struggles in the Garden of Olives with the will of the Father, it is not a matter of accepting for Himself a cruel disposition of God, but rather of attracting humanity into the very will of God.

We keep coming back to the mystery of the Trinity and the mystery of the Incarnation—mysteries we cannot fully explain. They keep haunting and challenging us. Yet they are the pillars of the Christian faith. Faith in God and in Jesus Christ is definitely more than the faith of reason alone. The best we can do is to make these mysteries more accessible to human reason as much as humanly possible. This reminds us again of what we discovered earlier—that, in the strictest sense, we can never entirely comprehend God.[201] We can try as much as is humanly possible, but after that, we can only become silent in awe and adoration.

---

[201] Thomas Aquinas, *Summa Theologica*, I, Q. 12, art. 7.

15

Is God the God of All Religions?

To answer the broad question posed in the title of this chapter, we first need to limit the scope of the terms *God* and *religion*—in other words, which God and which religions are we talking about? It's probably clear by now who the God we are talking about is: it is the God identified and defined by the proofs of God's existence, as was covered in the previous chapters. This God is Existence Itself, a Necessary Being, the First Cause, the Cosmic Designer, and an Eternal Intellect, all of which come with the attributes of being all-perfect, all-powerful, all-present, all-knowing, and all-good. Of course, God is more than that, but certainly not less than that.

Who is this God? Interestingly enough, when Moses asked God for His name, God answered him: "Yahweh," which means something like "I AM" or "I am who I am."[202] Jesus uses that very name for Himself when He said to the Jews, "Before Abraham came to be, I AM."[203] The name "I AM" expresses Pure Existence—the existence of a God who has no beginning and no end, a God who is absolute reality with no reality before Him and no reality outside of Him, a God who is utterly independent of anything else and on

---

[202] See Exod. 3:14.

[203] John 8:58 ("I AM" is in the present tense, which expresses eternity).

whom everything else depends. It is, in essence, the God to whom the proofs of God's existence refer.

This concept of God should be considered the bare "minimum" for anyone to be justified in speaking of God. Any god that doesn't meet this standard is not worthy to be called God but at best is an ideology, a superstition, an idol, or some other pseudo-god. All deities other than God are deceptions that lead to a quasi-religious adoration for deities other than God—the gods of the stars (astrology), the god of earthly possessions (mammon), the god of material goods (materialism), the goddess of nature (Mother Nature), the god of addictions, the idols of sex and food. To be completely oriented toward any of these—in short, to devote oneself unconditionally to anything less than God—is surrendering oneself to an idol. These so-called gods are so human and so little divine, so shallow and so little transcendent, that faith in them doesn't qualify as religious faith but, at best, as a make-believe faith.

Real religious faith, on the other hand, is about the framework in which "we live and move and have our being."[204] The Judeo-Christian tradition explicitly states that there are many gods that are called "idols," but there is only one real God. As we quoted earlier, "The LORD is our God, the LORD *alone!*"[205] And the book of Exodus repeats it: "You shall not have other gods beside me."[206] When we don't worship the real God, we must be worshipping idols instead, for we are by nature worshippers. What this leads to is well described by Chesterton with his witty pen: "For when we cease to worship God, we do not worship nothing, we worship anything."[207]

[204] Acts 17:28.
[205] Deut. 6:4 (emphasis added).
[206] Exod. 20:3.
[207] Quoted in Richard Keyes, "The Idol Factory," in Os Guinness and John Seel, eds., *No God but God* (Chicago: Moody Press, 1992), 32.

This also determines which faiths we can call religions. The distinction between God and idols eliminates already quite a few self-proclaimed religions from the list of real religions. There are some very weird ones. What about the Church of the Flying Spaghetti Monster, to name just one extreme case? In 2016, even a federal judge in the United States ruled that this church is not a real religion—let alone a *true* religion.[208]

Not all "religious beliefs" have a claim to our respect—we must respect their members, and perhaps even their beliefs, but not necessarily their truth claims. This might eliminate quite a few of them. Additional questionable candidates would be the Branch Davidians, Nuwaubianism, the Heaven's Gate, the Church of Scientology, the Nation of Yahweh, the Unification Church—and the list keeps growing, including even Satanism.

At one point, humanism wanted to be considered a religion. In its first manifesto (1933), it considered itself a religious movement to transcend and replace previous religions that were based on allegations of supernatural revelation. More recently, this was done so that humanism could be taught in public schools as a form of religious instruction. But that claim might work as a boomerang, because the First Amendment had created a wall of separation between religion and the state, forcing religions out of the public square, including public schools.

If humanism considers itself a religion, why is it allowed to be taught in public schools? It makes for a double standard: funding the teaching of secular humanism in public schools, while discriminating against traditional religions.[209] That is an inconsistency within secular

---

[208] Taylor Wofford, "Flying Spaghetti Monster Not Divine, Says Court," *Newsweek*, April 13, 2016, https://www.newsweek.com/flying-spaghetti-monster-parody-religion-447535.

[209] This conflict can be seen in the 1961 United States Supreme Court case *Torcaso v. Watkins*.

humanism itself: it likes to be a religion, and, at the same time, it intends to deny this title to any other traditional or competing sets of beliefs. But most of all, it doesn't acknowledge the existence of a higher being other than itself—which is essential for any kind of religion.

Other religions that should be eliminated from the list are the ones based on polytheism. As we found out earlier, God is that being in whom essence and existence are identical: God is pure Being and Existence. Hence, it is impossible for God to share an essence with anything else or to have one act of existence alongside others.[210] In other words, there is a logical proof for monotheism but not so for polytheism.

Nonetheless, we do find polytheism not only in Greek and Roman mythology but also in Chinese traditional religion, Hinduism, and Japanese Shintoism. Hindus, for instance, may focus their worship on one or more personal deities, while granting the existence of other gods, but the problem remains that these gods are not self-explanatory and therefore require a missing, ultimate First Cause to explain their existence. If Brahman is supposed to provide that source, then we may come closer to monotheism. But any belief in multiple deities defies all the proofs of God's existence.

Thus, our list of "legitimate" religions has basically been reduced to the three monotheistic religions: Judaism, Christianity, and Islam. What the Christian God, the Jewish God, and the Muslim God have in common is being all-perfect, all-powerful, all-present, all-knowing, and all-good. Those features combined make them at least a "legitimate" form of religion.

If you limit your list of what you call "legitimate" religions to the three monotheistic religions, you need to make sure that Christianity, with its triune God, deserves a place in that list.

---

[210] Thomas Aquinas, *Summa Theologica*, I, Q. 11, art. 3.

You are right; the triune and incarnate God is certainly not the God of the two other monotheistic religions. And what's more, some people think such a characterization seems to be in violation of God's attribute of being *one*. But that is so only at first sight. The oneness of God means only that God has no parts. Anything that is made up of parts depends for its existence on the one who has *no* parts but is perfect unity itself.

Just as the divine attributes are not parts of God, so are the three Persons in God not *parts* of God—each one is fully God. God the Father, God the Son, and God the Holy Spirit all refer to the same reality, which is the one and only God, although they are different "aspects" or "dimensions" of that reality—three different Persons, that is. We can think of our earlier example of talking about the "morning star" and the "evening star," where both expressions refer to the very same thing—the planet Venus. In a similar way, the three Persons in the Trinity refer to the same, fully unified reality—God. That's why Christianity is definitely a monotheistic religion, in spite of attacks made by the other two.

If "morning star" and "evening star" refer to the very same thing, why wouldn't the same apply to the Jewish God, the Christian God, and the Muslim God? I came to the conclusion that all three monotheistic religions are talking about the same God. Therefore, the Christian God must refer to the same God as the Jewish God and the Muslim God.

The problem of making this comparison is that "morning star" and "evening star" are both possible at the same time (not at the same time of the day, though). But it is impossible to claim that the Christian God is the same as the Jewish God or the Muslim God, because they contradict each other in many characteristics. When religions contradict each other, then at least one of them must be false. If they are all true, then none of them may be true.

Besides, a Muslim would never agree that the Muslim God is the same God as the Christian God—and neither would a Christian. The Christian God is a triune God, for instance, whereas the Muslim God is definitely not triune. They differ as much from each other as Aristotle's "sun" differs from Galileo's "sun." We can steer a spaceship to Galileo's sun but not to Aristotle's.

Perhaps a better analogy of their differences would be the following. It definitely matters whether we describe a whale as a *fish*, like a shark, or as a *mammal*, like a seal—the latter description is correct; the former is wrong. What both descriptions have in common is that they refer to vertebrates, but that's where their vertebrate similarities end. Unlike fish, whales are warm-blooded, they don't lay eggs, and they need to get oxygen from above the water. So we can't just take our pick—a whale is a mammal, not a fish. Treating a whale as a fish might be disastrous for the poor animal.

As fish and mammals have different characteristics, so do the Christian God, the Jewish God, and the Muslim God—we can't just gloss over their differences and declare them nonexistent. It does matter whether you believe in one or the others—they can't all be real and true at the same time, so at least two of them must be wrong. Again, what the Christian God, the Jewish God, and the Muslim God have in common, though, is being all-perfect, all-powerful, all-present, all-knowing, and all-good. But beyond that point, differences come in. As a result, your idea of a "common ground" between the three monotheistic religions begins to look more like thin ice.

Once we start talking about God, we begin to realize that our understandings of the divine can be very divergent—and that is why religions can be so different. Not all religions are of the same quality, however, no matter how tolerant you are about this issue. Those who think all religions are of the same quality basically defend some form of relativism in matters of religion—claiming that all religions are worth the same when seen within their own cultural setting. Ironically, however, relativism makes each religion absolute — its own authority in matters of truth. The idea that what is "true" in chess is also "true" in checkers, however, doesn't make sense. Besides, in this last analogy, we are merely dealing with man-made rules, but religion is about God-given truths.

The truth of the matter is what we said earlier: truth is truth, even if you do not accept it; and untruth is untruth, even if you claim it. Truth is truth — for everyone, anywhere, at any time. In other words, it needs to be stressed that we do not create truth. Truth needs to be discovered, not invented. As much as it is a matter of truth that the earth is not flat — believing that it is flat does not make it flat — it is also a matter of truth that God is tri-une — believing that He isn't cannot change God. Truth is truth, although we may not fully understand or capture the truth yet.

Even if we accept that religion must be about truth, we can still claim that all religions share in the same God-given truth, with each one having only part of that truth. Each religion may only have partial access to the truth — and that's why there is more than one religion.

You express here a rather common viewpoint. Popular in this context is the ancient Indian fable of six blind men who visit the palace of the raja and encounter an elephant for the first time. As each touches the animal with his hands, he feels a different part of the elephant, announcing an elephant to be all trunk, all tail, and so on. An argument ensues, each blind man thinking his own perception of the elephant is the correct one. The raja, awakened by the commotion, calls out from the balcony. "The elephant is a big animal," he says. "Each man touched only one part. You must put all the parts together to find out what an elephant is like."

The problem with this analogy is that the story presumes there *is* actually something like an elephant endowed with a trunk and a tail and so on, and that these men were just *blind* people, not able to "see" the real thing, an elephant. So the story assumes that there is a "real thing" that some may not see. The story assumes also that there is at least someone who knows the whole truth—the raja in this story—otherwise there is no way of talking about partial truths. The raja was in a position of privileged access to the truth. Because he could see clearly, he was able to correct those who were blind. If everyone is truly blind, then no one can know who is mistaken.

Nevertheless, the existence of so many religions is often used as an argument against any religion in particular. G.K. Chesterton was eager to debunk this argument with a simple analogy:

> It is perpetually said that because there are a hundred religions claiming to be true, it is therefore impossible that one of them should really be true.... It would be as reasonable to say that because some people thought the earth was flat, and others (rather less incorrectly) imagined it was round, and because anybody is free to say that it is triangular or hexagonal, or a rhomboid, therefore it has no shape at all; or its shape can never be discovered.[211]

In spite of what you and Chesterton say, it remains true that there is so much that different religions have in common. Any dialogue between religions is based on the idea of a "common ground," otherwise we couldn't even talk with each other about religion.

If we limit our discussion, for example, to the dialogue between Christianity and Islam, we soon discover that the common-ground approach obscures some very essential differences between the two. Both religions may refer to the same God in theory, but the way they

[211] G.K. Chesterton, "On Liberties and Lotteries," in *Collected Works* (Hastings, UK: Delphi Classics, 2014).

talk about this common reference is quite different. The title of a recent book raises a very fundamental question: "Is the Father of Jesus the God of Mohammed?"[212] Or are they essentially different?

Any kind of dialogue between religions must be honest about the differences that separate them—sparing us a false impression of common ground. In his encyclical *Caritas in Veritate*, Pope Benedict XVI expressed very clearly that the aim of any dialogue between religions is ultimately *truth*. The aim for both sides should be to come to the truth, albeit through respect and love. We can and should agree to disagree agreeably. We don't have to go along to get along! Those who think differently confuse respect for *people* with respect for *beliefs*. A deep respect for Muslims does not necessarily translate into having a deep respect for their beliefs.

The way we understand God and know Him penetrates everything else we do in life. If we say we love God, we need to *know* who He is. We cannot love what we do not know. Knowledge and truth are strongly connected. In fact, the concept of *truth* is at the core of all world religions, particularly in a religion such as Christianity. If Jesus never existed, the Word did not become flesh. If Jesus existed but was never crucified, then we were not redeemed. And as St. Paul wrote: "If there is no resurrection of the dead, then neither has Christ been raised. And if Christ has not been raised, then empty [too] is our preaching; empty, too, your faith."[213] Since truth is truth and untruth is untruth, religious beliefs are either true or untrue! Truth is of the essence.

This raises the question again as to how the Muslim God can be the same God that Christians adore. Yes, they both talk about "love of God and neighbor," but for Muslims this extends only

---

[212] Timothy George, *Is the Father of Jesus the God of Mohammed?* (Grand Rapids, MI: Zondervan, 2002).
[213] 1 Cor. 15:13–14.

to other Muslims. Human dignity in Islam comes from, and is conditional on, belief in and practice of Islam. The Muslim God commands that Jews and Christians should be subjugated or killed unless they accept the God of Islam. Allah curses anyone who says that God has a Son. Allah allows, or even promotes, a morality of polygamy. The Muslim heaven is a man's haven where each man is rewarded with seventy-two beautiful, high-bosomed virgins,[214] plentiful food, and slaves galore to attend to every whim and wish. This place is entirely different from the Christian Heaven, where, in the words of Roy Schoeman, himself a convert from Judaism to Catholicism, "the bliss comes from the pure joy of being in God's presence; in Islam it comes from base sensual pleasures—food, drink, and sex."[215] Are we really talking about the same God here?

In the previous chapter, you mentioned Arianism. The way you describe and identify a "real" religion, Arianism would qualify, too, to be on your list of real religions, I would say. It was a popular religion in the Roman Empire for a long time and definitely monotheistic.

Although Arianism is not a religion in the proper sense—but rather a heresy, or perhaps a sect—the case could be made that Arianism did in fact develop into a religion.

Arianism is not just something from the past. It is still alive in some form today—not only in Christianity in branches such as Unitarianism but also outside of Christianity in a religion such as Islam, which emerged some three centuries later, while Arianism was still very much alive and defended by many Roman emperors. We shouldn't forget that the Christians Mohammed knew from his part of the world were Arians, or had at least been affected by

---

[214] Qur'an 78:31–34.

[215] Roy Schoeman, *Salvation Is from the Jews* (San Francisco: Ignatius Press, 2003), 297.

Arian doctrines. The late historian Hilaire Belloc gave the following description of Islam:

> But the central point where this new heresy struck home with a mortal blow against Catholic tradition was a full denial of the Incarnation. Mohammed did not merely take the first steps toward that denial, as the Arians and their followers had done; he advanced a clear affirmation, full and complete, against the whole doctrine of an incarnate God. He taught that Our Lord was the greatest of all the prophets, but still only a prophet: a man like other men. He eliminated the Trinity altogether.... So true is this that today very few men, even among those who are highly instructed in history, recall the truth that Mohammedanism was essentially in its origins *not* a new religion, but a *heresy*.[216]

One of the first times we hear about the link between Islam and Arianism is in the writings of the Church Father John of Damascus (675–749), who lived shortly after Mohammed (who died in 632). In his work *Fountain of Knowledge*, he writes: "From that time to the present a false prophet named Mohammed has appeared in their midst. This man, after having chanced upon the Old and New Testaments and likewise, it seems, having conversed with an Arian monk, devised his own heresy."[217] John of Damascus was convinced that Islam was, in essence, not a new religion but a heretical form of Christianity. Mohammed did not learn much from Christianity, but he did get the gist of Arianism: Jesus is not God. Not surprisingly, the areas of the Roman Empire in which Arianism had been popular had now become wide open to Islam.

---

[216] Hilaire Belloc, *The Great Heresies*, chap. 4.
[217] *Fountain of Knowledge*, bk. 2, 101.

The case could even be made that Eastern Christian practice formed the template for what were to become the basic conventions of Islam. The Muslim form of prayer with its bowings and prostrations is strikingly similar to an older Syrian Orthodox tradition that is still practiced in some Christian churches across the Middle East. Those churches still have no pews, as in mosques. The architecture of the earliest minarets, which used to be square rather than round, unquestionably matches the church towers of Byzantine Syria. So it looks as if the Sufi Muslim tradition carried on directly from where the Christian Desert Fathers left off. The Ramadan—at first sight, one of the most authentic and alienating of Islamic practices—is, in fact, an Islamic version of Lent, which in some Eastern Christian churches still involves a strenuous all-day fast.

Besides, Islam contains several of the heretic elements we discussed earlier in connection with Arianism. Jesus was not crucified, it says, but only appeared to be crucified and then was taken up to heaven by Allah.[218] The doctrine of the "Trinity" is considered one of the worst blasphemies against the unity of Allah.[219] This Christian doctrine, as interpreted in the Qur'an, concerns instead a divine trinity of Allah, Jesus, and Mary.[220] Yet—in Muslim eyes—Jesus always denied that He was divine. To believe that Allah had a son is polytheism according to the Qur'an.[221] So one could very well make the case that Islam is not a new religion but rather a collection of old Arian heresies. Islam came after Christianity in time and then "corrected" Christianity's perceived errors, as older heresies had done before.

---

[218] Qur'an 4:157–8.
[219] Qur'an 4:48.
[220] Qur'an 5:116.
[221] Qur'an 4:171; 5:75; 9:30.

Can Christians, Jews, and Muslims still pray to the same God?

After all we have seen so far, it can no longer be said that all monotheistic religions worship the same God, for the way Christians understand God is not the same as the way Jews and Muslims understand God. Any kind of dialogue between religions has a tendency to stress the impression of common ground while ignoring the differences that separate them.

The 2011 interreligious prayer meeting in Assisi made unmistakably clear that gathering together to pray is not the same as praying together, for praying together implies that we are believing in the same God and are praying to the same God—which is not necessarily the case. On the other hand, people from different religions, although they may have very different understandings of the divine, can and should come together to pray *for* the same intention at the same time—such as their longing for peace—even though they cannot pray *to* the same God together.

In short, all of us can certainly pray together, but not to the same God. That's the tragedy of having different religions. Yet it's better to acknowledge reality than to just close our eyes to it. We owe it to God and to ourselves to strive for a religion that is *true*. Let's not forget Augustine's profound insight that "our heart remains restless until it rests in God"[222]—the true God, that is. False religions venerate false gods, which changes their gods into idols. They betray not only their own personal integrity but also, which is much worse, God's integrity.

---

[222] A paraphrase of Augustine, *Confessions*, 1, 1.

Conclusion

Can we prove that God exists? We found out that science does not have that power, if we want a proof in the sense of being certain and irrefutable. Do we have other ways to prove that God exists? Yes, we do. They are called proofs of God's existence, and they are based on a deductive form of reasoning that is logically conclusive.

These proofs start from a premise that is based not on specific religious insights but on a self-evident universal statement — one that is based not on scientific research or on a specific religion but on what we all share. Self-evident universal statements like these cannot be rejected without undermining anything we claim to know — they are the pillars or the basis of all our knowledge. Based on a premise with a self-evident universal statement, the argument leads us conclusively to God as Existence Itself, a Necessary Being, the First Cause, the Cosmic Designer, and an Eternal Intellect. The existence of this Being cannot be denied without rejecting the self-evident universal statement that the argument begins with.

Based on these proofs, we can logically derive that this Being must be unique, invisible, all-perfect, all-powerful, all-present, all-knowing, all-good, and all-loving. Rejecting the existence of this Being is first of all illogical. But if this Being does indeed exist, then denying its existence would also be a direct offense of God Himself.

Reason is able to prove this, all by itself. Is this the God most of us are familiar with? Not quite, so we found out. It is the God of reason, but not automatically the God of faith. The "living" God of faith is much richer than the "abstract" God of reason. We know more details about the living God of faith thanks to what God has revealed to us about Himself through His prophets, the authors of the Bible, and the Son of God, Jesus of Nazareth—collectively called special revelation. That's how we know that God is also a triune and incarnate God, in addition to what we know about God through reason alone. The unique God of reason and the triune God of faith may differ in sense, but they both refer to the same God, the one who is utterly unique.

The God of reason is shared by several religions, more in particular by the three monotheistic religions: Judaism, Christianity, and Islam (although they differ in stressing the role of reason). The God of reason may mostly be the same God for them, but the God of faith is not. A triune and incarnate God is definitely unique to only one of the three: Christianity. Does that mean that Christianity violates the God of reason? Not at all, so we found out. Christians have used reason to explain what they mean when speaking of a triune and incarnate God. By doing so, they have shown that it makes perfect sense to talk about God this way and that it can actually be the only true way to talk about God.

It is this living God of faith that is adored and revered in Catholicism. The danger is, though, that we may know much about God—the God of reason, that is, based on proofs of God's existence, for example—but without ever actually knowing God personally. St. Augustine says about the God of faith, "You were more inward to me than my most inward part and higher than my highest."[223] It requires an extra step to "know" God personally and

---

[223] Augustine, *Confessions*, 3, 6, 11.

intimately. In that sense, belief in the God of faith should also be a matter of "romance." St. Augustine is often quoted as the one who gave us this unusual perspective: "To fall in love with God is the greatest of romances, to seek him the greatest adventure, to find him the greatest human achievement."[224]

We have mentioned St. Thomas Aquinas numerous times in this book because of his brilliant mind when it comes to the two-some of faith and reason. But it should be mentioned also that this brilliant thinker left his master work, the *Summa Theologica*, unfinished because of an experience he had during Mass in December 1273. He told his secretary, Br. Reginald, "I can write no more. I have seen things that make my writings like straw."[225] He died a year later. On his deathbed, he humbly confessed to God: "Thee have I preached. Thee have I taught. Never have I said anything against Thee. If anything was not well said, that is to be attributed to my ignorance."[226] I would like to say the same about the contents of this book.

---

[224] I have never been able to locate this quote in Augustine's writings. Some say it appeared rather recently, with no direct connection to Augustine himself.

[225] Alban Butler, *Lives of the Saints*, rev. ed. (San Francisco: HarperSanFrancisco, 1991), 30.

[226] Dominic Prümmer, O.P., *Fontes Vitae S. Thomae Aquinatis, notis historicis et criticis illustrate* (Toulouse, 1912), 45.

Index

Index

redemption, 153
reductio ad absurdum, 30
reference, 84
regularity, 60
relativism, 168
revelation, 140
Russell, Bertrand, 16, 122
Ryle, Gilbert, 8

Sagan, Carl, 41
Sartre, Jean-Paul, 104
Sayers, Dorothy, 149, 159
Scalia, Antonin, 151
Schoeman, Roy, 172
Schrödinger, Erwin, 5
science, 3
self-explanatory, 30
self-reference, 122

sequence
    explanatory or hierarchical, 45
    linear or temporal, 44
Shaw, George Bernard, 57
space, 113
Suárez, Francisco, 12, 13

Tarski, Alfred, 75
teleology, 53, 59
Tertullian, 148
Thomas Aquinas. *See* Aquinas
time, 113
transcendence, 99, 101
Trinity, 167

verification, 16

watchmaker, 54
Wheeler, John A., 7

About the Author

Gerard M. Verschuuren is a human biologist, specializing in human genetics. He also holds a doctorate in the philosophy of science. He studied and worked at universities in Europe and in the United States. Currently semiretired, he spends most of his time as a writer, speaker, and consultant on the interface of science and religion, faith and reason.

Some of his most recent books are: *In the Beginning: A Catholic Scientist Explains How God Made Earth Our Home*; *Forty Anti-Catholic Lies: A Mythbusting Apologist Sets the Record Straight*; *Darwin's Philosophical Legacy: The Good and the Not-So-Good*; *God and Evolution?: Science Meets Faith*; *The Destiny of the Universe: In Pursuit of the Great Unknown*; *It's All in the Genes!: Really?*; *Life's Journey: A Guide from Conception to Growing Up, Growing Old, and Natural Death*; *Aquinas and Modern Science: A New Synthesis of Faith and Reason*; *Faith and Reason: The Cradle of Truth*; *The Myth of an Anti-Science Church: Galileo, Darwin, Teilhard, Hawking, Dawkins*; and *At the Dawn of Humanity: The First Humans*.

For more information, visit https://en.wikipedia.org/wiki/Gerard_Verschuuren.

Verschuuren can be contacted at www.where-do-we-come-from.com.

Sophia Institute

Sophia Institute is a nonprofit institution that seeks to nurture the spiritual, moral, and cultural life of souls and to spread the Gospel of Christ in conformity with the authentic teachings of the Roman Catholic Church.

Sophia Institute Press fulfills this mission by offering translations, reprints, and new publications that afford readers a rich source of the enduring wisdom of mankind.

Sophia Institute also operates the popular online resource CatholicExchange.com. *Catholic Exchange* provides world news from a Catholic perspective as well as daily devotionals and articles that will help readers to grow in holiness and live a life consistent with the teachings of the Church.

In 2013, Sophia Institute launched Sophia Institute for Teachers to renew and rebuild Catholic culture through service to Catholic education. With the goal of nurturing the spiritual, moral, and cultural life of souls, and an abiding respect for the role and work of teachers, we strive to provide materials and programs that are at once enlightening to the mind and ennobling to the heart; faithful and complete, as well as useful and practical.

Sophia Institute gratefully recognizes the Solidarity Association for preserving and encouraging the growth of our apostolate over the course of many years. Without their generous and timely support, this book would not be in your hands.

www.SophiaInstitute.com
www.CatholicExchange.com
www.SophiaInstituteforTeachers.org

Sophia Institute Press® is a registered trademark of Sophia Institute. Sophia Institute is a tax-exempt institution as defined by the Internal Revenue Code, Section 501(c)(3). Tax ID 22-2548708.